全国高等院校土木与建筑专业创新规划教材

# 土木工程专业英语
## English for Civil Engineering

俞家欢　主　编

清华大学出版社
北　京

## 内 容 简 介

本书是 21 世纪高等教育土木工程系列教材之一。全书由 16 个单元组成，每个单元包括课文精讲、翻译练习、材料泛读等内容。本书所选阅读材料涵盖土木工程总论、结构工程、岩土工程、道路与桥梁工程、隧道与地下工程、土木施工管理等方面，阅读材料内容新颖，大都是相关领域的新知识和新成果介绍。书中介绍的内容涉及专业英语的教学、语法特点、阅读方法及翻译方法与技巧等，对提高学生的专业英语阅读和翻译能力很有帮助。

本书可作为土木工程本科生专业英语课程的教材，也可作为土木工程研究生专业英语阅读的参考用书，还可供土木工程及相关领域的高校教师和工程技术人员阅读参考。

本书封面贴有清华大学出版社防伪标签，无标签者不得销售。
版权所有，侵权必究。举报：010-62782989，beiqinquan@tup.tsinghua.edu.cn。

图书在版编目(CIP)数据

土木工程专业英语/俞家欢主编．--北京：清华大学出版社，2017（2025.1重印）
（全国高等院校土木与建筑专业创新规划教材）
ISBN 978-7-302-46003-9

Ⅰ．①土… Ⅱ．①俞… Ⅲ．①土木工程—英语—高等学校—教材 Ⅳ．①TU

中国版本图书馆 CIP 数据核字(2016)第 316428 号

责任编辑：桑任松
装帧设计：刘孝琼
责任校对：周剑云
责任印制：杨 艳

出版发行：清华大学出版社
网　　址：https://www.tup.com.cn，https://www.wqxuetang.com
地　　址：北京清华大学学研大厦 A 座　　邮　编：100084
社 总 机：010-83470000　　邮　购：010-62786544
投稿与读者服务：010-62776969，c-service@tup.tsinghua.edu.cn
质量反馈：010-62772015，zhiliang@tup.tsinghua.edu.cn
课件下载：https://www.tup.com.cn，010-62791865

印 装 者：三河市天利华印刷装订有限公司
经　　销：全国新华书店
开　　本：185mm×260mm　　印 张：16.75　　字　数：403 千字
版　　次：2017 年 7 月第 1 版　　印　次：2025 年 1 月第 12 次印刷
定　　价：48.00 元

产品编号：067961-02

# 前　　言

近年来，随着我国对外开放的不断深化以及全球化浪潮的日益迅猛，我国高等院校学生的英语教学也面临着全新的要求，而专业英语是获取专业信息、掌握学科发展动态、参加国际学术交流的基本前提。专业英语教学必须以提高学生的英语综合应用能力为目标，使学生具备较强的跨文化交际能力，以应对未来职业的挑战。

《土木工程专业英语》作为高等院校专业英语系列教材之一，需要充分结合土木工程各专业的特点。

本书结合编者多年来土木工程专业英语课程的教学经验，在查阅了大量相关资料的基础上，在选材上做了认真的筛选，以注重专业基础内容和前沿专业知识为目标，本着覆盖面广、知识面宽以及适当介绍前沿专业知识的原则进行编写。本书是国内首本采用中英文完整对照形式的土木工程专业英语教材。每课所列生词表、专业术语、短语以及注释都是教学实践中学生经常提出的问题，因此针对性较强。此外，每课还安排了一定量的习题。读者通过对课文的阅读和练习，可以巩固专业英语的基础知识，扩大专业英语词汇和专业术语，从而具备一定的阅读专业英语文献的能力和翻译技巧，能够以英语为工具通过阅读去获取与本专业有关的国外前沿科技信息，了解本专业的最新国际动态。另外，在每章后还附有适量的习题供读者练习。将本书作为本科教材时，授课学时可采用32学时，教师可根据实际情况对讲授内容进行取舍或增删。

本书课文和阅读材料语言规范，题材广泛，覆盖土木工程各专业的重要内容。全书共分16章：土木工程研究领域，迪拜——在建中的城市，土木工程材料，荷载，结构设计，结构建筑发展简史，钢筋混凝土，钢结构，砌体结构，木结构，组合结构，道路设计，桥，测量的基本概念，防灾与减灾工程，项目规划、组织及施工控制。课文的难易程度切实结合本科学生的实际水平。本书重视语言技能训练，突出对阅读和翻译能力的培养，以求达到《大学英语专业阅读阶段教学基本要求》所提出的目标："通过指导学生阅读有关专业的英语书刊和文献，使他们进一步提高阅读和翻译科技资料的能力，并能以英语为工具获取专业所需的信息。"

英语知识覆盖面广、专业英语教学与实际的有机结合以及文献阅读选材广泛，是本书的三个主要特色。编者希望通过学习本书，能使土木工程专业的学生较全面地掌握专业英语知识和应用技能，为今后的工作打下良好的基础；也希望土木科技工作者能借助本书使自己的专业英语知识得到更新、充实和提高，对实际工作有所裨益。

承蒙清华大学出版社约稿，值此书付梓之际，编者向清华大学出版社、关心和支持本书编写的同事以及参考文献的作者表示诚挚的谢意和敬意。由于编者水平有限，书中不妥之处望广大读者提出宝贵意见。

编　者

## 本书编委会

主　编：俞家欢

参　编：于　玲　金　路　闫林伟　白晓彤

　　　　罗光友　刘永年　杨千荨　李姗姗

　　　　金子辉　阚　强

# 目 录

## Chapter 1　Scope of Civil Engineering
## 第 1 章　土木工程研究领域 ........................................................................................... 1

1.1　Civil Engineering　土木工程 ................................................................................ 1

1.2　Role of Civil Engineers　土木工程师的作用 ....................................................... 4

1.3　Impact of infrastructural Developments on the National Economy　基础设施的开发对
　　 国民经济的影响 ..................................................................................................... 6

New Words and Expressions　生词和短语 ..................................................................... 7

Exercises　练习 ................................................................................................................ 7

Answers　答案 .................................................................................................................. 8

## Chapter 2　Dubai — A City under Construction
## 第 2 章　迪拜——在建中的城市 ................................................................................. 10

2.1　Structural Dubai　迪拜建筑 ............................................................................... 10

2.2　Integrated Design　整体设计 ............................................................................. 13

2.3　Transportation　交通体系 ................................................................................. 14

2.4　Environmental Dubai　环境建设 ....................................................................... 18

2.5　Water Resources　水资源 .................................................................................. 20

New Words and Expressions　生词和短语 ................................................................... 22

Exercises　练习 .............................................................................................................. 23

Answers　答案 ................................................................................................................ 24

## Chapter 3　Civil Engineering Materials
## 第 3 章　土木工程材料 ................................................................................................. 26

3.1　Traditional Materials　传统建筑材料 ............................................................... 28

3.2　Composite Materials　复合材料 ........................................................................ 36

3.3　Smart Materials　智能材料 ............................................................................... 38

New Words and Expressions　生词和短语 ................................................................... 40

Exercises　练习 .............................................................................................................. 41

Answers　答案 ................................................................................................................ 42

## Chapter 4　Loads
## 第 4 章　荷载 ................................................................................................................. 44

4.1　Vertical Loads　竖向荷载 .................................................................................. 44

|  |  |  |
|---|---|---|
| 4.2 | Lateral Loads　水平荷载 | 48 |
| 4.3 | Other Loads　其他性质的荷载 | 54 |
| New Words and Expressions　生词和短语 | | 56 |
| Exercises　练习 | | 57 |
| Answers　答案 | | 58 |

## Chapter 5　Structure Design
## 第 5 章　结构设计 ........................................................................................................ 60

|  |  |  |
|---|---|---|
| 5.1 | Science and Technology　科学与技术 | 60 |
| 5.2 | Structural Engineering　结构工程 | 60 |
| 5.3 | Structures and their Surroundings　结构及环境因素 | 61 |
| 5.4 | Architecture and Engineering　建筑与结构 | 61 |
| 5.5 | Architectural Design Process　建筑设计过程 | 62 |
| 5.6 | Architectural Design　建筑设计 | 62 |
| 5.7 | Structural Analysis　结构分析 | 63 |
| 5.8 | Structural Design　结构设计 | 63 |
| 5.9 | Load Transfer Mechanisms　荷载传递机理 | 64 |
| 5.10 | Structure Types　结构承重体系 | 65 |
| New Words and Expressions　生词和短语 | | 67 |
| Exercises　练习 | | 68 |
| Answers　答案 | | 69 |

## Chapter 6　A Brief History of Structural Architecture
## 第 6 章　结构建筑发展简史 ............................................................................................ 71

|  |  |  |
|---|---|---|
| 6.1 | Before the Greeks　古希腊时期 | 71 |
| 6.2 | Greeks　希腊时期 | 72 |
| 6.3 | Romans　罗马时期 | 74 |
| 6.4 | The Medieval Period (477-1492)　中世纪时期(477—1492) | 75 |
| 6.5 | The Renaissance　文艺复兴时期 | 77 |
|  | 6.5.1　Leonardo da Vinci　列奥纳多·达·芬奇 | 78 |
|  | 6.5.2　Brunelleschi　布鲁内莱斯基 | 78 |
|  | 6.5.3　Galileo　伽利略 | 79 |
| 6.6 | Pre Modern Period, Seventeenth Century　新世纪早期，十七世纪时期 | 81 |
|  | 6.6.1　Hooke　胡克 | 81 |
|  | 6.6.2　Newton　牛顿 | 82 |
| 6.7 | The Pre-Modern Period: Coulomb and Navier　新世纪早期：库仑，纳维 | 84 |
| 6.8 | The Modern Period (1857 to Present)　新世纪时期(1857 年至今) | 86 |
| New Words and Expressions　生词和短语 | | 88 |

Exercises　练习 ............................................................................................................. 89

　　　Answers　答案 ............................................................................................................... 90

## Chapter 7　Reinforced Concrete Architecture
## 第 7 章　钢筋混凝土 ........................................................................................................... 92

　　7.1　Use in Construction　在建筑中的应用 ..................................................................... 93

　　7.2　Behavior of Reinforced Concrete　钢筋混凝土的性能 ............................................ 93

　　7.3　Reinforcement and Terminology of Beams　梁的增强和概念 ................................. 95

　　7.4　Common Failure Modes of Steel Reinforced Concrete　钢筋混凝土常见的破坏形式 ......... 97

　　7.5　Carbonation　碳化作用 ............................................................................................. 97

　　7.6　Chlorides　氯化物 ..................................................................................................... 98

　　　New Words and Expressions　生词和短语 ................................................................. 100

　　　Exercises　练习 ........................................................................................................... 101

　　　Answers　答案 ............................................................................................................. 102

## Chapter 8　Steel Structure
## 第 8 章　钢结构 ................................................................................................................. 104

　　8.1　History of Steel Structures　钢结构的发展历史 ................................................... 104

　　8.2　Steel Structures Characteristic　钢结构的特点 ..................................................... 106

　　8.3　Application Of Steel Structure　钢结构的应用 ..................................................... 108

　　8.4　Aim of Steel Structural Design　钢结构的设计目标 ............................................ 112

　　8.5　Limit State　极限状态 ............................................................................................ 113

　　8.6　Load and Calculation of Load Effects　荷载与荷载效应计算 ............................. 114

　　8.7　Material Selection　材料选择 ................................................................................ 115

　　8.8　Design Indices　强度设计值 .................................................................................. 116

　　8.9　Provisions for Deformation of Structures and Structural Members　结构及构件变形限值 ......... 117

　　　New Words and Expressions　生词和短语 ................................................................. 119

　　　Exercises　练习 ........................................................................................................... 121

　　　Answers　答案 ............................................................................................................. 122

## Chapter 9　Masonry Structure
## 第 9 章　砌体结构 ............................................................................................................. 124

　　9.1　Introduction　引言 .................................................................................................. 125

　　9.2　Materials to Use　材料的使用 ................................................................................ 125

　　9.3　Elements are Part of The Structural System　结构系统的组成 ............................ 128

　　9.4　Preparation before Starting the Construction　施工前准备 ................................. 129

　　9.5　How to Build the Foundation　基础施工 .............................................................. 129

　　9.6　How to Build the Over Footing　地梁施工 ........................................................... 132

9.7　How to Build A Wall　墙体施工 .................................................................................. 132

9.8　How to set the Confining Columns to the Wall　构造柱施工 ...................................... 135

9.9　How to Build the Slab and Beams　梁、板施工 .......................................................... 136

9.10　How to Finish the Surface of the Elements　构件表面做法 ....................................... 138

New Words and Expressions　生词和短语 ............................................................................ 139

Exercises　练习 ...................................................................................................................... 140

Answers　答案 ....................................................................................................................... 141

## Chapter 10　Building with Wood
## 第 10 章　木结构 ............................................................................................................... 143

10.1　Wood Frame Construction: Low-rise Solutions　木框架结构：低层建筑 ................ 144

10.2　Wood Frame in China　木结构在中国的发展 ............................................................ 146

10.3　Hybrid Construction: Wood Frame Storeys on Concrete Structure　混合建筑：木框架与混凝土结构结合 ................................................................................................. 147

10.4　Engineered Wood Construction: Solid Wood Panels　工程木结构：实木板 ............ 149

10.5　Engineered Wood Construction: Glued Laminated Timber　工程木结构：木料胶合板 ............ 151

New Words and Expressions　生词和短语 ............................................................................ 153

Exercises　练习 ...................................................................................................................... 155

Answers　答案 ....................................................................................................................... 156

## Chapter 11　Composite Structures
## 第 11 章　组合结构 ........................................................................................................... 157

11.1　Applications of Concrete-filled Steel Tubes　钢管混凝土的应用 ............................. 157

11.2　Advantages of Concrete-filled Steel Tubes　钢管混凝土的优点 ............................... 163

New Words and Expressions　生词和短语 ............................................................................ 164

Exercises　练习 ...................................................................................................................... 165

Answers　答案 ....................................................................................................................... 166

## Chapter 12　Pavement Design
## 第 12 章　道路设计 ........................................................................................................... 168

12.1　Introduction　引言 ...................................................................................................... 168

12.2　Flexible Pavement　柔性路面 .................................................................................... 169

12.3　Rigid Pavement　刚性路面 ........................................................................................ 174

Exercises　练习 ...................................................................................................................... 178

Answers　答案 ....................................................................................................................... 179

## Chapter 13　Bridge
## 第 13 章　桥 ....................................................................................................................... 181

13.1　Introduction　引言 ...................................................................................................... 181

| 13.2 | Beam Bridges 梁式桥 | 181 |
| 13.3 | Cantilever Bridges 悬臂桥 | 183 |
| 13.4 | Arch Bridges 拱桥 | 185 |
| 13.5 | Suspension Bridges 悬索桥 | 188 |
| 13.6 | Cable-stayed Bridges 斜拉桥 | 190 |
| 13.7 | Truss Bridges 桁架桥 | 191 |

New Words and Expressions 生词和短语 ... 192
Exercises 练习 ... 194
Answers 答案 ... 195

## Chapter 14　Basic Concepts in Surveying
## 第 14 章　测量的基本概念 ... 197

| 14.1 | Introduction 引言 | 197 |
| 14.2 | Classification 测量类型 | 198 |
| 14.3 | Principles of Surveying 测量原理 | 200 |
| 14.4 | Modern Surveying Instruments 现代测量仪器 | 201 |
| 14.5 | GIS　GIS 系统 | 206 |
| 14.6 | Remote Sensing and Their Applications 遥感技术及其应用 | 208 |

New Words and Expressions 生词和短语 ... 210
Exercises 练习 ... 211
Answers 答案 ... 212

## Chapter 15　Disaster Prevention and Reduction
## 第 15 章　防灾与减灾工程 ... 214

| 15.1 | Principal Causes of Disasters 灾害的主要成因 | 214 |
| 15.2 | Earthquake 地震 | 215 |
| 15.3 | Geological Disasters: Landslide, Collapse and Debris Flow 地质灾害：滑坡，崩塌和泥石流 | 217 |
| 15.4 | Rainstorm and Flooding 暴雨和洪水 | 225 |
| 15.5 | Disaster Self Rescue 灾难发生后的自救 | 228 |
| 15.6 | Some Major Effects of Disasters 一些主要的灾害影响 | 229 |
| 15.7 | Disaster Plan 灾难应变计划 | 230 |

New Words and Expressions 生词和短语 ... 234
Exercises 练习 ... 235
Answers 答案 ... 236

## Chapter 16　Planning, Scheduling and Construction Management
## 第 16 章　项目规划、组织及施工控制 ..................................................................... 238

16.1　Construction Management　施工管理 ........................................................ 238
16.2　Organizing for Project Management　项目管理组织 ................................. 242
16.3　Defining Work Tasks　规定工作任务 ......................................................... 249
New Words and Expressions　生词和短语 ......................................................... 252
Exercises　练习 .................................................................................................. 253
Answers　答案 ................................................................................................... 254

## 参考答案 ................................................................................................................. 256

# Chapter 1　Scope of Civil Engineering
# 第1章　土木工程研究领域

## 1.1　Civil Engineering　土木工程

"Civil engineering is the profession in which a knowledge of the mathematical and physical sciences gained by study, experience, and practice is applied with judgment to develop ways to utilize, economically, the materials and forces of nature for the progressive well-being of humanity in creating, improving, and protecting the environment, in providing facilities for community living, industry and transportation, and in providing structures for the use of humanity."

——American Society of Civil Engineers, 1961

"土木工程是通过从研究、经验及实践中获得的数学、物理等自然科学知识，结合自身的经验判断较为经济地利用材料、自然界的力量，来开发、利用及保护环境，并为人类的生活、生产及交通及居住提供便利条件。"

——美国土木工程协会(1961)

To date, more than 200 projects worldwide have earned this prominent designation, which illustrates the creativity and innovative spirit of civil engineers. Almost always performed under challenging conditions, each of these engineering feats represents the achievement of what was considered an impossible dream.

时至今日，世界上超过 200 个土木工程领域的项目被公认为优秀设计，其中的每个项目都代表了土木工程师的创造性及创新精神。并且几乎每个项目都是在充满挑战性的条件下完成，每项工程都曾被认为是一个不可能实现的梦想。

Civil Engineering is the oldest branch of Engineering next to Military Engineering. It involves planning, design, construction and maintenance of structures such as bridges, roads, canals, dams, tunnels and multi-storied buildings. Shelter is the basic need of mankind's existence. The huts built with bamboos and leaves can be taken as the early Civil Engineering Constructions carried out to satisfy the needs for shelter (as shown in Figure 1.1). Over the years there has been a tremendous growth in the field of civil engineering to provide quality houses which are safe, functional, aesthetic and economical.

土木工程是仅次于军事工程的最古老的工程学科之一。它包含诸多工程结构，如桥梁、道路、河道、隧道及多层建筑等的规划、设计、施工及维护等多方面的内容。挡风遮雨是人类生存的最基本需求。利用竹子与树叶搭建起来的简易帐篷是人类最早的建筑(如图 1.1

所示)。一直以来，土木工程的发展目标是为人类提供安全、适用、美观及经济的高质量的住所。

Figure 1.1  Hut
图 1.1  木屋

Flats and apartments are constructed in urban areas to provide shelter to a large number of people. Multi-storied buildings and skyscrapers(as shown in Figure 1.2) are planned and constructed to provide office spaces, shopping malls, cyber centers, hotels, restaurants, and etc.

公寓及住宅多建于人口密度大的城市以满足大量人口的居住要求。因此，多层建筑及摩天大楼(如图 1.2 所示)被设计并建造出来，为人类提供办公、购物、网吧、酒店及餐馆等场所。

Figure 1.2  Multi-storied Buildings and Skyscrapers
图 1.2  多层房屋及摩天大楼

Irrigation is defined as the artificial application of water to land for the purpose of raising crops. Civil Engineering gives vast scope for irrigation by constructing barrages, dams (as shown in Figure 1.3), canals and distributaries. Vast areas of dry land have been successfully irrigated and green revolution has become a reality in China.

灌溉是利用人工将水引到土里，为庄稼提供必要的水分。土木工程涉及大量的水利方面的建设，如水库、大坝(如图 1.3 所示)、运河及引流工程等。大面积的干旱地区得以被成功灌溉，从而使绿色产业在中国成为现实。

Figure 1.3　Dam
图 1.3　大坝

Providing better transportation facilities is also a part of civil Engineering. Good network of roads, highways and expressways are necessary for movement of men and materials. Railways, airways and waterways (as shown in Figure 1.4) are needed for overall development of any country.

提供便利的交通运输条件也是土木工程的一个重要功能。优化的路网建设对劳动力及材料的运输来说是必不可少的。铁路、航空及水运(如图 1.4 所示)对一个国家的全面发展至关重要。

Figure 1.4　Railways, Airways and Roads
图 1.4　铁路、机场及城市道路

Water is an important need for all living beings. Potable water supply to the towns and cities is an important link in Civil Engineering. Natural water available in lakes and rivers is not suitable for drinking without proper treatment. Hence big water treatment plants have been constructed and operated to supply potable water to the public. In addition, waste water generated in the city has to be taken out and disposed after providing suitable treatment. Waste water treatment plants (as shown in Figure 1.5) are constructed to prevent pollution of surface and ground water sources.

Flood control and environmental protection are other areas in Civil Engineering which has an important role to play. Dams and levees are constructed to mitigate floods.

对所有生命个体来讲，水是必不可少的。对城镇居民供水是土木工程领域十分重要的环节。湖泊及江河中自然状态的水未经适当的处理是不适合人类直接使用的。因此，大量的水处理厂不断被兴建并投入使用，为公众提供优质供水。另外，城市中产生的污水在排放前也必须经过适当的处理，污水处理厂(如图 1.5 所示)可防止地表水及地下水资源污染。

防洪及环境保护是土木工程的另一重要功能。大坝及码头可用来减轻洪水的影响。

Figure 1.5　Water Treatment Plant

图 1.5　污水处理厂

## 1.2　Role of Civil Engineers　土木工程师的作用

A Civil Engineer is one who deals with the planning, designing, construction and maintenance of different types of civil engineering work. He should be competent in various fields such as surveying, analyzing, estimating, construction scheduling and construction management.

土木工程师处理各种类型的土木工程项目的规划、设计、施工及维护工作，擅长不同领域，如勘察、分析、预估、施工进度及施工管理等方面的工作。

A Civil Engineer will involve in various engineering activities such as:

土木工程师参与如下各种工程活动，如：

(1) Surveying (as shown in Figure 1.6) and preparation of estimates;

工程勘察(如图 1.6 所示)和估算准备；

Figure 1.6　Surveying

图 1.6　工程勘察

(2) Planning, designing and construction of houses, apartments, office - buildings, commercial establishments and factory buildings;

房屋、住宅、办公建筑、商业建筑及工业建筑的规划、设计及施工；

(3) Planning and design of transportation facilities such as highways and railways;

规划、设计交通工程，如高速公路、铁路等；

(4) Construction of ports and harbors, railway stations, bus and truck terminals, airports and helipads (as shown in Figure 1.7);

港口、铁路站台、公交站、航空港及停机坪(如图1.7所示)等的建设；

Figure 1.7　Airports and Helipads
图 1.7　机场及直升机停机坪

(5) Construction of dams and canals (as shown in Figure 1.8) is for irrigation and drinking water supply for flood control purposes; planning, design and construction of pollution control facilities such as sewage treatment plants.

水利工程，如大坝、运河(如图1.8所示)及饮用水工程等施工；污染控制设施，如污水处理厂等的规划、设计及施工。

Figure 1.8　Dam and Canal
图 1.8　水坝与运河

## 1.3 Impact of Infrastructural Developments on the National Economy 基础设施的开发对国民经济的影响

The term infrastructure is used to denote the conditions which are available for economic development of a region. In other words, infrastructure can be defined as the facilities to be provided by the state or central government or local administration for overall development of a region, which include power generation, transportation, health, education, water and sanitation and other public utilities(as shown figure 1.9).

基础设施通常是衡量一个国家(地区)经济发展重要指标。换言之，基础设施是中央政府或地方政府为国家或地区的全局性发展所提供的条件，包括发电、交通运输、健康、教育、饮水和卫生以及其他公共事业(如图1.9所示)。

The investments made on infrastructural developments have a profound impact on the national economy. Some of the effects of investment in infrastructure are given below.

在基础设施方面的投入对国家的发展具有深远的战略意义，具体内容如下。

Investments in infrastructural facilities result in increased job opportunities for both skilled and unskilled, literate and illiterate people. Construction of new roads, bridges and canals provide employment to large number of people. The creation of better infrastructure in a region motivates entrepreneurs to establish their own industries, service centers, commercial establishments etc. This will open up job opportunities to a number of unemployed people in the region. The improvements in job opportunities will have positive effects on economy.

基础设施建设可以提供更多的就业机会，不论就业人口是否是文盲或者有无技术，新建道路、桥梁、运河等的施工不可避免地需要大量的人工。一个地区好的基础设施可以激励企业家建立自己的产业、服务中心、商业网点等，这些均会对本地区大量的无业人员的再就业提供机会。较高的就业程度对经济发展有积极的作用。

Figure 1.9　Industry Facility
图1.9　工业基础设施

## New Words and Expressions　生词和短语

1. civil engineering　土木工程
2. progressive　*adj.* 进步的；先进的
3. creativity　*n.* 创造力；创造性
4. innovative　*adj.* 革新的，创新的
5. maintenance　*n.* 维护，维修；保持
6. canal　*n.* 运河
7. tunnel　*n.* 隧道；坑道；洞穴通道
8. multi-storied　*adj.* 多层的
9. bamboo　*n.* 竹，竹子
10. aesthetic　*adj.* 美的；美学的；审美的
11. skyscraper　*n.* 摩天楼，超高层大楼；特别高的东西
12. irrigation　*n.* 灌溉
13. barrage　*n.* 火力网，弹幕射击；阻塞；齐射式攻击；[水利] 拦河坝
14. facility　*n.* 设施；设备
15. highway　*n.* 公路，大路
16. expressway　*n.* (美)高速公路
17. harbor　*vt.* 庇护；藏匿；入港停泊
　　　　*n.* 海港；避难所
18. treatment　*n.* 处理；对待
19. levee　*n.* 堤坝(码头)
20. sewage treatment plant　污水处理厂
21. infrastructural　*adj.* 基础结构的

## Exercises　练习

Ⅰ. Write a T in front of a statement if it is true according to the text and write an F if it is false.

1. Civil Engineering is one branch of Military Engineering.
2. Irrigation is defined as the artificial application of water to land for the purpose of increasing crops.
3. Civil Engineer can deal with planning, designing, construction, demolishing and maintenance of different types of Civil Engineering work.
4. Infrastructure can be defined as the facilities to be provided by the government or local administration.

Ⅱ. Complete the following sentences.

1. Civil engineering is the profession in which a knowledge of the_____ and _____ sciences gained by study, experience, and practice is applied with judgment to develop ways to utilize, economically, the materials and forces of nature for the progressive well-being of humanity in _____, _____, and _____the environment, in providing facilities for _____, _____and _____, and in providing structures for the use of humanity.

2. Civil Engineering involves _____, _____, _____and _____of structures.

3. _____, _____ and _____ are needed for overall development of any country.

4. A Civil Engineer should be competent in various fields such as _____, _____, _____, _____, _____.

5. The infrastructures include _____, _____, _____, _____, _____, _____.

Ⅲ. Translate the following sentences into Chinese.

1. The huts built with bamboos and leaves can be taken as the early civil engineering constructions carried out to satisfy the needs for shelter.

2. Natural water available in lakes and rivers is not suitable for drinking without proper treatment. Hence big water treatment plants have been constructed and operated to supply potable water to the public.

3. The term infrastructure is used to denote the conditions which are available for economic development of a region.

4. The creation of better infrastructure in a region motivate entrepreneurs to establish their own industries, service centers, commercial establishments, and etc.

5. A Civil Engineer is one who deals with the planning, designing, construction and maintenance of different types of civil engineering works.

Ⅳ. Translate the following sentences into English.

1. 土木工程是工程学中仅次于军事工程学的最古老的分支。
2. 公寓建在城市地区是为了给大批量人口提供住所。
3. 灌溉的定义是利用人工把水应用到土地达到种植作物的目的。
4. 投资基础设施的发展对这个国家的经济有深远的影响。
5. 这些设施包括电力、交通、健康、教育、供水、卫生和其他公共事业。

# Answers 答案

Ⅰ.

1. F  2. T  3. F  4. T

Ⅱ.

1. mathematical, physical, creating, improving, protecting, community living, industry, transportation

2．planning, design, construction, maintenance

3．railways, airways, waterways

4．surveying, analyzing, estimating, construction scheduling , construction management

5．power generation, transportation, health, education, water, sanitation

Ⅲ.

1．用竹子和树叶建造的小屋可以作为早期土木工程建设推行的满足需要的住房。

2．未经适当处理的湖泊和河流中的自然水不适合饮用，因此大的水处理厂已建成并运营，将饮用水供应给公众。

3．长期基础设施是为一个地区的经济发展提供可用的条件。

4．一个地区更好的基础设施的创造激励着企业家们建立自己的产业、服务中心、商业机构等。

5．一个土木工程师是一个可以处理规划、设计、施工和维护等不同类型土木工程工作的人。

Ⅳ.

1．Civil Engineering is the oldest branch of Engineering next to Military Engineering.

2．Flats and apartments are constructed in urban areas to provide shelter to a large number of people.

3．Irrigation is defined as the artificial application of water to land for the purpose of raising crops.

4．The investments in infrastructural developments have a profound impact on the economy of the country.

5．The facilities include power generation, transportation, health, education, water and sanitation and other public utilities.

# Chapter 2　Dubai — A City under Construction
# 第 2 章　迪拜——在建中的城市

　　Dubai, United Arab Emirates is a city experiencing dramatic and swift urbanization, the result of which includes many groundbreaking works of civil engineering. The reinvestment of petroleum revenue into infrastructure has led to many challenging feats of engineering, from the tallest building in the world, the most expensive resort islands in the world, to the world's largest airport. No matter how groundbreaking and exciting these developments are, they come at a significant cost to the environment and water resources of this desert country. The advances in the fields of structural, geotechnical, transportation and water resource engineering, as well as the associated efforts to keep them in check with responsible and sustainable development, are part of the positive goal of working to keep civil engineering innovation and profitability alive while respecting the environment and the needs of future generations.

　　阿拉伯联合酋长国的迪拜是一个正在经历戏剧性快速城市化的城市，其结果是不断出现各种各样的开创性建筑结构。从世界第一高楼、世界最昂贵的人工岛到世界最大的航空港，石油财政对基础设施的再投资使得迪拜产生了很多富于特点的标志性建筑。然而这样对刺激性发展对这个沙漠地区的国家的环境与水资源造成了严重的影响。在结构、地质、交通、水资源工程等领域的进展以及为使它们符合负责任和可持续发展而做的相关工作都促使土木工程行业有生机、健康地发展，与此同时还要确保环境及子孙后代的发展需要。

## 2.1　Structural Dubai　迪拜建筑

　　Dubai, a city prominent for its large scale construction boom, is making breakthroughs in the field of structural engineering. With its fast paced and expensive construction projects, Dubai is quickly setting new records for the tallest building and hotel, and the largest waterfront, thus popularizing the commercial aspect of the city and increasing tourism. With the creation of these record breaking structures, Dubai is incorporating the latest technologies and designs to form, for example, its tallest building, an underwater hotel, and a rotating building.

　　迪拜，一个正在大兴土木的城市，在建筑结构方面不断地创造奇迹。因其快速的发展及造价高昂的工程项目不断上马，迪拜在超高层建筑、宾馆、滨海区的建设上不断地突破纪录，从而给这座城市带来了更大的商业吸引力及繁荣的旅游业。正因如此，迪拜正在吸收最先进的技术及设计来建造世界上最高的建筑、一座水下宾馆及一座旋转建筑等。

Dubai's change from an oil-based to a tourist directed economy is seen through its record breaking commercial construction projects. Originally, its economy was based on the natural resources of its region, oil and natural gas, but tourism and property now make up most of its revenue. This can be seen from the amount of expansion undertaken by the state owned Dubai World and its real estate development company, Nakheel, that have created projects such as the artificial Palm Islands and the World, with an interest in beating the world records. The Burj Dubai, the now tallest building, exceeds the previous record holder, the 509 m tall Taipei 101 and holds residential apartments, a 5+ star Giorgio Armani Hotel, and corporate offices. A comparison of the Burj Dubai and other skyscrapers is as shown in Figure 2.1.

迪拜从一个以石油为基础的城市发展成为当今以旅游业为主要经济增长点的城市，其发展可视为由其以商业开发为主的工程项目不断突破的过程。最初，其经济主要依靠地区性的自然资源，如石油、天然气等，但如今财政收入主要依靠旅游业。这一点可从由国家控股的"迪拜世界"及其房地产公司 Nakheel(棕榈岛集团)的扩张看出，它们曾成功地开发了人工棕榈岛项目，其收益打败了世界纪录。迪拜目前最高的建筑迪拜塔，其高度已超过中国台北 509 米高的 101 大厦，内有公寓住宅、一家超五星级的乔治·阿玛尼酒店以及公司办公室。如图 2.1 所示为迪拜塔与其他摩天大楼的对比。

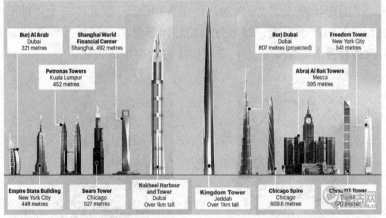

Figure 2.1  The Burj Dubai and Several of the World's Tallest Buildings

图 2.1  迪拜塔及世界的最高建筑物

Dubai also holds the world's largest mall, containing many of the world's largest clothing lines, numerous hotels, cinemas and restaurants. A major tourist attraction is the Burj Al Arab, the world's second largest building used solely as a hotel. Situated on an artificial island, the hotel is among the most expensive in the world, with the per night stay cost ranging from $1,000 to $28,000. Another major construction project in Dubai is the Hydropolis Underwater Hotel, 20 m below the Persian Gulf, with 220 suites with a $6,000 per night cost. The common theme in all these projects is that in some way, they are each bigger or taller than similar renderings around the world. By creating the most extravagant, unique, and architecturally pleasing structures of commercial use, Dubai is attracting large populations of people who come to work in or tour the state.

迪拜拥有世界上最大的商场，其中有多条世界上最大的服装制造生产线，数不清的旅馆、电影院及饭店。最主要的标志性建筑是迪拜阿拉伯酒店，是世界上纯粹用于宾馆的第二大建筑，是坐落在人工岛的世界上最昂贵的酒店，每晚消费 1000 至 28 000 美元。另一个主要的项目是迪拜海底酒店(其名称为 Hydropolis，音译为海德帕利斯)，位于波斯湾水下 20 米深处，有 220 个每晚消费 6000 美元的房间。迪拜的其他项目在某种程度上都较其他国家的同类建筑要大要高！通过建造这种奢侈的、独一无二的、富于建筑艺术气息的商用建筑，迪拜正不断地吸引全世界的人来此工作或旅游。

The construction of commercial buildings has in part been expensive and in other parts a great advancement in civil engineering. Take the Burj Dubai for example, in aiming for the tallest building, one of the main factors to be considered was wind resistance. For this reason, the Y-shaped buttressed design, as shown in Figure 2.2, was applied to the design foundation. With the three different sides, the building does not severely face weather effects from any single side, as each of the three wings balance the intake of any outside force. The tripod foundation is also a strong base for the building, as it maximizes the area of ground coverage.

以商业为目的开发的建筑在某种程度上比较昂贵，但另一方面也会促进土木工程的发展。以迪拜塔为例：其目的是成为世界上最高的建筑，施工中需考虑的一个因素即为风的阻力。基于此，平面形状设计成"Y"形，如图 2.2 所示。因此"三个外伸部分"组成的整体结构不会受到天气的严重影响，每一方向的外伸部分均会平衡掉外界荷载的作用。"三角架"式的基础具有强大的承受能力，并可以使占地面积最大化。

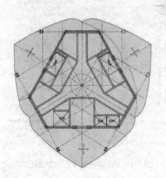 

Figure 2.2　Typical Floor Plan of the Burj Dubai
图 2.2　迪拜塔平面图

## 2.2 Integrated Design  整体设计

Another major project that is implementing the latest technologies to its grandiose design is Hydropolis, the first underwater luxury hotel. The hotel is to be made 20m deep near the Palm Islands of Dubai in the Persian Gulf. The creation of an underwater structure that will actively be in use requires a lot of attention to detail, because it deals with people underwater, and also deals with the underwater environment, habitats of underwater creatures, and the rates and patterns of tides and currents. But the reasonability for such construction projects is questionable. So much money is put into the projects, resulting in the service being unattainably expensive. The Hydropolis Hotel required two years to gather $500 million since the idea of the large scale underwater project appeared questionable to investors. The expense of the construction is so high that right now it has reached a halt. What the hotel will look like upon completion is as shown in Figure 2.3.

另一个主要的项目迪拜海底酒店也采用了当前最先进的技术来满足其浮夸的设计，即人类史上首个最奢华的水下宾馆，它位于波斯海湾人工岛水面下 20 米深度处。水下建筑的设计需要考虑若干因素，因为它使人生活于水下，需考虑水下环境、水下生物的生存习惯及潮起潮落的速率。但无论如何建造这样的水下建筑都是具有争议的。投入的经费如此巨大，使得服务价格非常昂贵。该水下宾馆需用两年的时间收入 5 亿美元来吸引投资者。如此昂贵的造价已达到顶峰。如图 2.3 所示为该水下旅馆建成之后的效果图。

Figure 2.3  Hydropolis Underwater Hotel, Dubai, United Arab Emirates
图 2.3  迪拜海底酒店

Recent reports show that Dubai is currently under $8bn of debt, and has also postponed payment of a $3.5bn bond. But more projects are still being conceived and under construction. One major project that is raising a lot of interest is the Dynamic Tower. Architect David Fischer's 1,380 foot tall skyscraper will have levels that will constantly be changing at different speeds and will be 130 ft taller than the Empire State Building. The building will change by the forces of nature; as the wind turbines that turn the floors allow for their rotation, it will also generate the electricity to run the building. The innovation of this design is immense, but the $6,000 per square foot price of the structure is questionable for a state in recession. Different buildings in Dubai are as shown in Figure 2.4.

有报道表明，目前迪拜已有 80 亿美元的债务，已延迟支付 35 亿美元的工程款。但目前仍有多项建设项目正在规划待建。一个吸引了人们极大的兴趣的项目即为动力大厦。由建筑师大卫·费思特设计的 1380 英尺高的摩天大楼，能够不断地改变运动速度，比纽约的帝国大厦高出 130 英尺。该建筑将通过自然界的力量为风力涡轮机旋转楼层提供动力，其产生的电能也将用来运行整栋建筑。这种创新是史无前例的，但每平方英尺 6000 美元的造价使这个正处于经济萧条的国家备受争议。如图 2.4 所示为迪拜的各种建筑。

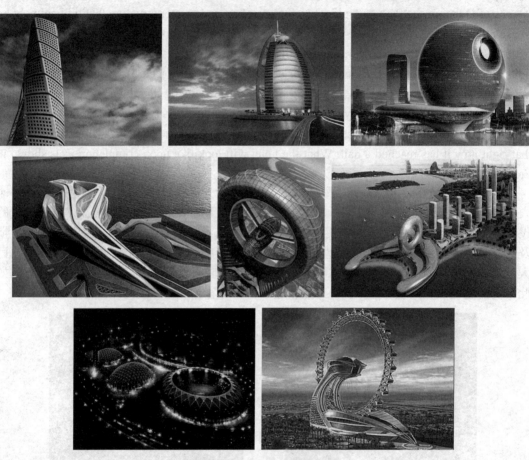

Figure 2.4　The Buildings of Dubai

图 2.4　迪拜的奇特建筑

## 2.3　Transportation　交通体系

As with any area of large-scale urban expansion, Dubai has encountered a demand for transportation services. These services are required to fill two important voids—the need for a reliable and affordable intra-city transportation infrastructure for the 2 million local residents, and a luxurious inter-city system for the tourism industry that has dominated the local economy. This pressing need, combined with recent economic woes and an uncertain future regarding billions of dollars of debt, has forced a consolidation of resources that leaves the transportation sector of the local economy experiencing troubling "growing pains."

在大规模城市化的背景下，迪拜对交通设施的需求日益增大。基础交通设施的建立能够填补两大空白：一是为当地 200 万居民的出行提供可靠的保证；二是为占本国主导经济地位的旅游业提供非常便利的交通体系。在这种背景下需要将目前经济困境与即将可能产生的亿元债务通盘考虑，减轻当地交通设施的建设所带来的"阵痛"。

Traffic congestion, a problem faced by many urban areas, has been particularly troublesome for the Dubai area. Rapid growth has created a high demand for vehicle transportation, and the subsequent congestion and delays causes an estimated Dh4.6 billion (U.S. $1.25 billion) per year, which amounts to more than 3% of the GDP. The Roads and Transport Authority(RTA), the governing body in charge of regulation, planning, and development, has implanted several policies and plans in an attempt to fix this problem. Dr. Abdul Malek Ebrahim Abu Shaikh, head of the RTA's Transportation Studies and Planning group, cites the high car ownership rate as a cause of the problem. Dubai's rate of 541 cars per 1,000 people exceeds that of New York, London, and Singapore (444, 345 and 111 respectively). Regulating the amount of driver's licenses issued, increasing the vehicle tax, and implementing tolls are all either being evaluated as solutions or in the early stages of implementation.

交通拥挤这一很多城市所面临的问题对迪拜来说十分突出。城市快速的发展要求高效的交通设施。随之而来的交通拥挤带来了保守估计每年约 12.5 亿美元的损失，超过了 GDP 的 3%！道路交通管理部门是执行管理、规划及发展职能的政府机关，采取了多项措施来解决这个问题。阿卜杜勒·马利克·易卜拉欣·阿布·谢赫博士，交通管理及规划研究所负责人，指出导致交通拥堵的根本原因在于个人拥有小汽车的数量太多所致。在迪拜，每千人拥有 541 辆汽车，甚至高于纽约、伦敦及新加坡(分别为 444 辆、345 辆及 111 辆)。限制驾驶证的发放数量、提高机动车税收及通行费用可能是解决问题的途径。

An important cause of the traffic congestion problem is the financial crisis that has impacted most of the world since 2008. Mortgage foreclosure and escalating rent costs have driven many Dubai residents to seek cheaper living in distant locations. Public transportations (the Dubai Metro and bus systems) do not yet reach these areas and are forcing thousands to make a longer daily commute. In addition to the increase in distance travelled, many experts say the region is "suffering from an originally flawed road system, with built-in bottlenecks on certain key routes such as the Dubai-Sharjah road".

交通拥挤的一个重要因素是自 2008 年以来的席卷全球经济危机的影响。抵押物回赎权丧失及租金的上涨迫使迪拜本地居民到偏远的地方寻找更便宜的住所。公共交通还远未到达这些地方，造成了居民每天长距离的往返。很多专家认为除了上下班距离远外，迪拜最初交通设施的错误规划也是很重要的一个方面，在某些关键路段易形成"瓶颈"，如迪拜—沙迦公路。

Even with a large-scale switch from private to public transportation, many of the problems and inefficiencies would not be alleviated due to issues with funding allocation and, most recently, debt consolidation as the region's tourism-dependent income has been on the decline. A Gulf Talent survey of more than 5,000 professionals in the area concludes that "a core underlying problem remains that, across the region, the development of support infrastructure is lagging behind more prestigious mega-projects such as airports, business parks, and high-rise towers – leading to continuous bottlenecks and disruptions in traffic."

即便对私有及公共交通部分做出了调整，但由于财政投入及近期债务重组的问题，这一地区以旅游业为主的收入下降，很多问题及效率低下仍未得到解决。一项对超过 5000 个专业人士的海湾人才调查表明：根本的问题是迪拜支持性基础设施的投入远未达到或远滞后于某些有声望的大型项目，诸如机场、商业园区及超高层建筑等，进一步加剧了城市发展的"瓶颈"问题及交通的混乱。

During the height of the luxurious urbanization, excessively high amounts of money were dedicated to ensuring that Dubai's image on the world map would be one of modern luxury that rivaled Las Vegas and Monte Carlo. In the process, however, requisite components of the transportation infrastructure were either under-funded or completely neglected.

在高水平的城市化建设进程中，迪拜投入了大量的财力来保证迪拜的现代奢华的"世界印象"，以媲美拉斯维加斯和蒙特卡罗。这个过程中，交通的基础设施建设获得的投入很少，甚至完全被忽略。

Construction of the Dh 12.45 billion (U.S. $3.4 billion) project began in 2005 under the directive of Sheikh Mohammed bin Rashid Al Maktoum (the Prime Minister and Vice President of the United Arab Emirates). Two troubling aspects of the Metro (as shown in Figure 2.5) are cause for immediate concern and will likely impair the effectiveness of the project. First, the system was planned primarily to service the luxurious and iconic structures located on the coast and city skyline.

投资 34.5 亿美元、由阿拉伯联合酋长国总理谢赫·穆罕默德·本·拉希德·阿勒马克图姆亲自主持的地铁项目(如图 2.5 所示)于 2005 年开工建设。该项目有两个令人担心的问题亟须关注并且有可能对项目的实施产生损害。首先，该项目立项的初衷主要是服务于位于海岸及城市天际线的奢华及有代表性的建筑。

Figure 2.5　Dubai Metro Line
图 2.5　迪拜的地铁

Peyman Parham Younis of the RTA reinforces this purpose when he states: "When you talk about Dubai you talk about the seven-star hotel — the Burj AI Arab — or the Palm Island —the first man-made island — or the tallest building in the world — the Burj Dubai. We want the Metro to become a new icon and to connect all of these icons". While transportation between the "iconic" buildings and financial districts in Dubai is certainly important, neglecting to build adequate routes between them and the poorer residential areas which house thousands of service industry employees does little to alleviate traffic concerns. Second, the Metro (with the exception of corporate sponsorships paid for station naming rights) was underwritten by the Dubai Government. Dubai World, the "emirate's investment arm," is (U.S.) $59 billion dollars in debt and is now facing pressure from international banks and other creditors regarding lavish spending that may not generate enough income.

项目负责人佩曼·巴翰·尤尼斯这样陈述其建设目的："当你谈论迪拜时，你会谈到七星级宾馆——迪拜帆船酒店、第一个人工岛及世界最高的建筑——迪拜塔等，我们也想使地铁成为一个新的元素来实现上述标志性建筑间的联系。"虽然迪拜的各标志性建筑及金融区之间的交通相当重要，但忽略了在这些地方和安置了成千上万服务业雇员的、较贫困的住宅区之间修建充足的道路，使得在缓解交通方面收效甚微。其次，地铁项目(除了为车站命名权支付的企业赞助费外)主要由政府负责。迪拜世界，"埃米尔投资公司"负债 590 亿美元，目前主要的还贷压力来源于国际银行及其他信贷机构，其投资巨大，可能不会带来太多的收益。

While the Dubai World debts themselves are not directly guaranteed by the Dubai government, the two are heavily intertwined due to Dubai World's heavy investments on the luxurious coastline. The Metro, RTA, and many other important infrastructure components will be impacted by this crisis for years to come due to the fact that they are all almost completely government owned and operated. It seems as though the worst is yet to come, as budget cuts and project cancellations will almost certainly follow.

由于迪拜政府对迪拜世界的债务不给予直接担保，迪拜世界在豪华海岸线上的过大投入使得二者紧密地联系在一起。地铁项目道路交通及其他项目会因此在以后数年内受到极大影响，因为这些项目几乎均由政府控制和运营。似乎最坏的事情尚未发生，但随着预算的削减，某些项目几乎肯定会被取消。

The Metro project serves as a viable microcosm for the problems faced by the city as a whole. "Gold Class" service with "leather seats and plush pile carpets offers panoramic views at the front of the train." Coach passengers must settle for only air conditioning, Wi-Fi, and cell phone reception. Had luxury been sacrificed for effectiveness and availability, could the RTA have funded additional lines to reach the poorer neighborhoods of the city? Upon full completion of the project (which may be years away), the average upper-class financial analyst will have an easier commute from their high-rise condo to the financial district. The waiters, bartenders, and thousands of other lower-class service industry personnel, however, may not benefit as well when their bus remains at a standstill in rush hour traffic.

地铁可视为城市作为一个整体所面临的问题的缩影，这从在能提供全景的火车上铺上长毛绒地毯并设置真皮座椅这种"黄金"级服务可以看出。旅客车厢必须布设空调、无线网络及手机信号接收装置。这种奢华的设施是否具有有效性及实用性？RTA 能否建设可以到达城市里贫穷的街区的线路？若此工程完工(可能要数年后)，普通的上层社会的金融分析师可十分容易地往返于他们的高层公寓与办公区之间；而侍应生、酒保及成千上万服务行业的工薪阶层由于所乘坐的公交车在交通高峰期间停滞不前,行驶缓慢,可能并没有从中受益。

As Dubai moves into an uncertain future, the transportation sector requires more emphasis on practicality, which should come at the cost of luxury. The fate of this unique and spectacular region may depend on the Dubai Government's ability to focus on broad and vital concepts and to invest their resources in an effective transportation infrastructure.

由于迪拜正进入一个未知发展的阶段,交通设施建设要求提出日程,其实施也会按大投资的奢华模式建设。这个独一无二的、特殊的地区的命运可能取决于迪拜政府是否有能力、有眼光及在有效的交通基础设施建设上投入的资源。

## 2.4 Environmental Dubai 环境建设

The mass development of Dubai has contributed significantly to the current global warming trend. Global warming is about the temperature increase in the earth, and the corresponding rise in greenhouse gasses as a percent of the total atmospheric composition.

迪拜的过度发展加剧了全球变暖问题。全球变暖是指地球表面温度升高,使得大气层内某些温室气体成分相应地升高。

This is caused by green house gases such as carbon dioxide, methane, nitrous oxide, ozone, and CFCs entering the atmosphere of the Earth. In Dubai, the rate of green house gases emission has rapidly increased due to the recent mass constructions. According to Dubai and Green engineering, the seawater desalination in Dubai emits 1.6 times more $CO_2$, one of the major greenhouse gases, than in California. Eventually, with the world's increasing rate of green house gas emission, the sea level will increase such that the artificial palm island, shown in Figure 2.6, could be underwater in 50 years.

这主要是由于温室气体,如二氧化碳、沼气、一氧化二氮、臭氧及氯氟烃进入了大气层所致。迪拜近期的大面积施工使得温室气体的含量快速提高。根据迪拜的情况及环保部门的研究:迪拜的海水淡化释放出更多的二氧化碳(一种主要的温室气体),是加利福尼亚的1.6倍。最终,随着温室气体的不断排放,海平面将会不断上升,如图2.6所示的人工岛在50年后将会低于水面。

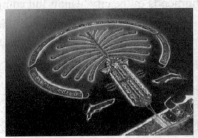 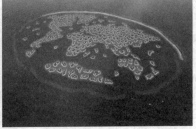

Figure 2.6　Dubai Artificial Island

图2.6　迪拜人工岛

Dubai faces many challenges to the environment as a result of the mega projects. The Palm Jumeirah, which is an artificial island built into the Persian Gulf in Dubai has a major role in destroying its nearby ecosystem. According to a research project from the World Environmental and Water Resources Congress 2009, the artificial islands are being built without proper environmental studies, and as a result the Persian Gulf's eco-system is in danger. The islands damage the local aquatic environment, and cause rapid erosion of the sand. Specifically the coastal beaches lost more than 3,500,000 m$^3$ of sand per year. That is a big difference when it's compared to the historical lost of 10,000 – 15,000 m$^3$ per year. Also, the artificial islands in Dubai face many natural hazards such as earthquakes, tsunamis, and tropical cyclones. The islands are in danger of earthquakes because of liquefaction and slope failure. Earthquakes from the Iranian Costal Source Zone can cause the building on the islands to be significantly damaged.

迪拜因其超大项目的不断建设面临着许多环境方面的问题。建于波斯湾的人工岛——朱美拉棕榈岛严重地破坏了近海生态。世界环境组织及水资源协会的一份研究报告表明：人工岛在建设时没有进行正确的环境评估，结果导致波斯湾的生态系统遭到破坏。人工岛破坏了当地的海洋生态环境，加剧了当地沙漠化进程。尤其是海岸线每年流失350万立方米的沙子。这与历史记载的每年流失10 000~15 000立方米有很大的差别。同时人工岛还面临其他的自然灾害，如地震、海啸及台风的影响。人工岛极易受到地震的影响，因其易于液化及发生断裂破坏。源自伊朗海岸源区的地震危险会使人工岛上的建筑受到严重的威胁。

Tsunamis can also be dangerous to the islands. Even though the Arabian Gulf is an inland sea which should ostensibly be safe from tidal waves, history shows that the region can be affected by tsunamis. In 1945, when the Makran earthquake occurred, the tsunami created by the earthquake affected the Arabian Gulf region. If this type of tsunami occurs again, the islands can be flooded since it's in the ocean.

海啸也会对人工岛产生危害。即便阿拉伯海湾为内陆海，似乎不会受到潮汐的影响，但有史料记载也曾受到海啸的破坏。1945年莫克兰(Makran)地震引起的海啸对该地区造成了影响，若类似的海啸再度发生，因人工岛处于海洋中，势必会受其影响。

When humans develop a city, they must consider the environmental consequences because they can affect people lives via their impacts on the eco-system. The environmental issues in Dubai, global warming, pollutant discharge, and natural hazards must be resolved to avoid drastic consequences. Dubai administrators need to conduct environmental impact assessments (EIA), which are "the most widely practiced environmental management tool". With EIA, they can avoid negative the environmental consequences while continuing to develop their economy and infrastructure.

人类进行城市建设时，必须考虑对环境的影响，因为它们会通过作用于生态系统而对人们的生活产生影响。迪拜的环境问题，如全球变暖、污染排放及其他自然灾害必须加以控制以免造成更严重的后果。迪拜政府需要运行目前在世界范围内广泛应用的环境评估机制(EIA)。通过这一机制，可以避免其经济及基础设施的发展过程中产生的负面环境影响。

## 2.5　Water Resources　水资源

　　As a byproduct of the oil production and downstream distribution which peaked in the late 80's, Dubai has undergone an economic restructuring which has dramatically impacted water demand in the emirate. As Dubai has transitioned from a fishing and trading village, to an oil production and distribution center, and finally to a diversified economy based on real estate, the population has leapt from 52,000 to 2,260,000 between 1968 and 2008. This dramatic increase in growth, as well as a substantial increase of industrial usage of water has pushed Dubai's water resource needs beyond their already meager groundwater's ability to fulfill.

　　作为石油行业的副产品，在其产业链下游的分布于 20 世纪 80 年代达到顶峰，迪拜正在进行的经济重组已经对酋长国水资源的需求产生了很大的影响。迪拜由一个以渔业及贸易为主的村庄发展为石油生产及配送中心，再到现在已形成了以房地产开发为主的多元化经济实体，其人口由 1968 年的 5.2 万人增加到 2008 年的 226 万人。城市的快速增长，以及工业用水的急剧增加，使得迪拜对水资源的需求已远超其贫乏的地下水储量！

　　As Dubai has coped with this challenge, and continued its incredible growth, it has had to innovate and diversify its water resources to meet these demands. A great deal of the current water needs of Dubai have been met through desalination of seawater. As a port city, Dubai has never had a shortage of seawater, and they have used several forms of cutting edge desalination technology to utilize this resource. This increase in potable water production from sea water has been resource intensive, expensive, and comparatively wasteful.

　　由于迪拜已面临如此的挑战，同时又需进行难以置信的发展，必须采取创新措施来丰富其水资源以满足其需要。目前迪拜所需的大量的水来自海水的淡化。作为一个港口城市，迪拜从不缺乏海水，也采取了各种措施来淡化海水。但这种获得可用水的方法造价高昂并且相当浪费！

　　For the purposes of this examination and engineering critique, I will examine the construction of Dubai's largest water desalination plant, the Jebel Ali plant (as shown in Figure 2.7). This plant is a combined power generation (natural gas) and desalination (multiple flash distillation columns) plant, to be finished in July 2010 at a projected cost of 550 million US dollars. Combined power generation and desalination means that the desalination process is piggybacked on the excess thermal energy from the power generation. By coupling these processes together, engineers are able to make the desalination cheaper and more efficient, as well as reduce the pollutant stream.

　　以迪拜最大的海水淡化厂——杰贝阿里海水淡化厂(如图 2.7 所示)的建设为例说明。该厂既可进行海水淡化也可生产天然气，于 2010 年 7 月完工，总投资 55 亿美元。因海水的淡化是通过大量的热能而完成的，设计师将两者统一起来是为了降低海水淡化成本及减少污染的排放量。

Figure 2.7　Jebel Ali Desalination Plant
图 2.7　杰贝阿里海水淡化厂

In the case of flash distillation desalination plants, the main point-source pollutant is referred to as brine. Brine is the heated and concentrated saltwater which is the waste product of desalination. While this may not intuitively feel like a serious contaminant, the EPA regards it as industrial waste, and for good reason. When brine is dumped back into an eco-system, the higher dissolved solid content, and the corresponding high density of the liquid makes it sink to the bottom of the body of water. Once there, the salt content (usually twice the salinity of normal sea water), causes all living organisms to undergo extreme osmotic pressure. This cellular pressure literally tears microorganisms such as plankton apart, destroying the most plentiful food source in the ocean. Additionally, the higher temperature causes larger marine species to seek new habitats.

在这个庞大的海水淡化厂，主要的污染物是卤水(浓盐水)，它是海水加热后的沉淀物。因其可能在直观上不会给人严重污染物的感觉，环境保护协会视其为工业废物，这不无道理。一旦卤水进入生态系统，吸收的固体物质越多，相应地卤水的浓度越高，会使其沉入海底。若含盐量达到正常海水的两倍时，会使得海洋生物承受极大的渗透压力。这种渗透力会破坏微生物诸如浮游生物，破坏海洋中大量的食物来源。另外，较高的温度会使大量的海洋生物寻求新的习性。

The costs associated with this means of production are great, and normally this would fundamentally alter the economic reality of water usage by both industry and individual citizens. However, in the emirate of Dubai, water and electricity prices are set by a central authority. As a result of these practices by the government utility provider (Dubai Electricity & Water Authority, DEWA), Dubai's residents are actually amongst the most wasteful in the world, with a per capita daily domestic water usage of 550 liters. For comparison, the United States daily per capita domestic water usage is between 200 and 300 liters, depending on region.

The pricing of water in Dubai is on a graduated scale depending on consumption levels, which was first instituted in 2006 as a reaction to the loss of money and energy that results from the massive desalination of seawater. Currently this applies to all non-Emirates, which accounts for 83% of the population and can be seen as a comparison with prices locally in Blacksburg and averages across the United States.

海水淡化的经济代价过高，一般来讲会从根本上增加民用和工业用水的费用。但是在迪拜，水与电的价格完全由中央政府确定，由政府公用事业提供商(迪拜水电局，DEWA)

进行运作的结果是，实际上，世界上最浪费水资源的人口中迪拜的居民是其中之一，每家每天用水 550 升。而与之对比的美国每家日均用水量在 200～300 升。

迪拜的水价主要按用量实行阶梯水价，这项标准是于 2006 年实施的，主要是应对因过量的海水淡化造成的费用及能源消耗过大。目前这项规定主要是对约占总人口 83%的非埃米尔居民适用，可与当地的布莱克斯堡居民美国平均数相对比。

Clearly Dubai pays a price for its rapid development and the infrastructure necessary to support the growth. While the necessary steps are clearly possible, and are currently being implemented, the question is if the long-term costs are sustainable. Only time will tell, but it is clear that paying several times as much for as ubiquitous a commodity as water puts great economic strain on the Emirate of Dubai.

显然，迪拜正在为其城市化的快速发展及支撑这一发展所必需的基础设施而付出代价。显然在发展过程中必要的措施是可行的，并且近来不断被落实，问题是长时间的投入是否具有可持续性。相信只有时间会告诉我们，但对水资源的投入程度再加大些，也将会对迪拜的经济发展带来新的压力。

## New Words and Expressions　生词和短语

1. urbanization　　*n.* 都市化；文雅化
2. groundbreaking　*n.* 动工
　　　　　　　　　*adj.* 开创性的
3. waterfront　*n.* 滩，海滨；水边
4. rotate　*vt.* 使旋转；使转动；使轮流
5. artificial island　人工岛
6. Y-shaped buttressed design　Y 形的支撑设计
7. intake　*n.* 摄取量；通风口；引入口
8. integrate　*n.* 一体化；集成体
　　　　　　　*adj.* 整合的；完全的
9. grandiose　*adj.* 宏伟的；宏大的
10. traffic congestion　交通堵塞
11. bottleneck　*n.* 瓶颈；障碍物
12. the United Arab Emirates　阿拉伯联合酋长国
13. iconic　*adj.* 图标的，形象的
14. wastewater contamination　废水污染
15. methane　*n.* [有化] 甲烷；[能源] 沼气
16. nitrous oxide　一氧化二氮；笑气(等于 laughing gas)
17. ozone　*n.* [化学] 臭氧；新鲜的空气
18. desalination　*n.* 脱盐作用；减少盐分
19. ecosystem　*n.* 生态系统
20. liquefaction　*n.* 液化；熔解

21. tsunami    *n.* 海啸
22. marine species    海洋生物

# Exercises  练习

Ⅰ. Write a T in front of a statement if it is true according to the text and write an F if it is false.

1. Dubai has made breakthroughs in the field of structural engineering.
2. Dubai has the world's largest mall, including many of the world's largest clothing lines, numerous hotels, cinemas, and restaurants.
3. Due to issues with funding allocation, even with a large-scale switch from private to public transportation, many of the problems and inefficiencies would be alleviated.
4. The Metro is $59 billion dollars in debt and is now facing pressure from international banks and other creditors regarding lavish spending that may not generate enough income.
5. The Arabian Gulf is an inland sea which should ostensibly be safe from tidal waves, and history also shows that the region cannot be affected by tsunamis.
6. When brine is dumped back into an ecosystem, the higher dissolved solid content, and the corresponding high density of the liquid makes it sink to the bottom of the body of water.

Ⅱ. Complete the following sentences.

1. An important cause of the traffic congestion problem is _____.
2. Those massive projects have dramatically affected the _____, _____, and _____ of the Dubai.
3. Green house gases include _____,_____,_____,_____,_____.
4. Dubai maybe face many natural hazards,such as _____,_____,_____.
5. The environmental issues in Dubai include _____,_____,_____.
6. In the emirate of Dubai, _____ and _____ are set by a central authority.

Ⅲ. Translate the following sentences into Chinese.

1. The reinvestment of petroleum revenue into infrastructure has led to many challenging feats of engineering, from the tallest building in the world, the most expensive resort islands in the world, to the world's largest airport.
2. Dubai, a city prominent for its large scale construction boom, is making breakthroughs in the field of structural engineering.
3. Dubai's change from an oil-based to a tourist directed economy is seen through its record breaking commercial construction projects.
4. As with any area of large-scale urban expansion, Dubai has encountered a demand for transportation services.
5. Traffic congestion, a problem faced by many urban areas, has been particularly

troublesome for the Dubai area.

Ⅳ. Translate the following sentences into English.

1. 交通拥挤的一个重要的原因是金融危机从2008年已经影响到世界上大部分地区。

2. 许多其他重要的基础设施组件将在数年内受到这一危机的影响，因为它们几乎完全由政府拥有和经营。

3. 这是由于温室气体引起的，如进入地球大气层的二氧化碳、甲烷、一氧化二氮、臭氧和氟氯烃。

4. 当人类发展了一个城市，他们必须考虑环境后果，因为环境可以通过影响生态系统影响人们的生活。

5. 很显然，迪拜将为其迅速发展以及支撑其发展的必要的基础设施付出代价。

# Answers　答案

Ⅰ.

1. T  2. T  3. F  4. T  5. F  6. T

Ⅱ.

1. financial crisis
2. economy, environment, demographics
3. carbon dioxide, methane, nitrous oxide, ozone, CFC
4. earthquakes, tsunamis, tropical cyclones
5. global warming, pollutant discharge, natural hazards
6. water, electricity prices

Ⅲ.

1. 从世界第一高楼，世界最昂贵的人工岛到世界最大的航空港，石油财政的基础设施的再投资使得迪拜产生了很多富于特点的标志建筑。

2. 迪拜，一个正在大兴土木时期的城市，在结构工程方面不断地创造奇迹。

3. 迪拜由以石油为基础向以旅游业为经济主导的转变，其发展可视为由其以商业开发为主的工程项目不断突破的过程。

4. 在大范围城市化的背景下，迪拜对交通设施的需求日益增大。

5. 交通拥挤是很多城市所面临的问题，这一问题对迪拜来说尤其突出。

Ⅳ.

1. An important cause of the traffic congestion problem is the financial crisis that has impacted most of the world since 2008.

2. Many other important infrastructure components will be impacted by this crisis for years to come due to the fact that they are all almost completely government owned and operated.

3. This is caused by green house gases such as carbon dioxide, methane, nitrous oxide, ozone, and CFCs entering the atmosphere of the Earth.

4. When humans develop a city, they must consider the environmental consequences because they can affect people lives via their impacts on the eco-system.

5. Clearly Dubai pays a price for its rapid development and the infrastructure necessary to support the growth.

# Chapter 3　Civil Engineering Materials
# 第 3 章　土木工程材料

Several materials are required for construction. The materials used in the construction of Engineering Structures such as buildings, bridges and roads are called Engineering Materials or Building Materials. They include Bricks, Timber, Cement, Steel and Plastics. The materials used in Civil Engineering constructions can be studied under the following headings.

工程建设需要使用多种材料。用于工程结构，诸如建筑、桥梁及道路施工的材料称工程材料或建筑材料，包括砖、木、水泥、钢材及塑料等。在下面的标题中，我们会研究土木工程建筑中常用的材料。

It is necessary for an engineer to be conversant with the properties of engineering materials. Right selection of materials can be made for a construction activity only when material properties are fully understood. Some of the most important properties of building materials are grouped as shown in Table 3.1.

土木工程师必须熟悉工程材料的性能。只有充分掌握材料的性能，经准确地选材而形成的构件才可用于施工过程。建筑材料的一些重要的特性总结如表 3.1 所示。

The general classification of building materials is as follows:

建筑材料的大体分类如下：

(a) Traditional materials;

传统材料；

(b) Alternate building materials;

可替代的建筑材料；

(c) Composite materials;

复合材料；

(d) Smart materials.

智能材料。

Table 3.1　Some Properties of Building Materials

| Group | Properties |
| --- | --- |
| Physical | Shape, Size, Density, Specific Gravity etc. |
| Mechanical | Strength, Elasticity, Plasticity, Hardness, Toughness, Ductility, Brittleness, Creep, Stiffness, Fatigue, Impact, Strength etc. |
| Thermal | Thermal conductivity, Thermal resistivity, Thermal capacity etc. |
| Chemical | Corrosion resistance, Chemical composition, Acidity, Alkalinity etc. |

Continued 续表

| Group | Properties |
|---|---|
| Optical | Colour, Light reflection, Light transmission etc. |
| Acoustical | Sound absorption, Transmission and Reflection |
| Physiochemical | Shrinkage and Swell due to moisture changes |

表 3.1 建筑材料的性能

| 类　别 | 性　能 |
|---|---|
| 物理特性 | 形状、尺寸、重度、特殊的比重等 |
| 力学特性 | 强度、弹性、塑性、硬度、延性、脆性、徐变、刚度、疲劳、冲击强度等 |
| 热力学特性 | 热传导、热阻系数、热容量 |
| 化学特性 | 耐腐蚀性、化学组成、酸碱度 |
| 光学特性 | 色、光反射、光传播等 |
| 声学特性 | 声吸收、传播及反射 |
| 生化特性 | 因潮湿因素等引起的收缩、膨胀等 |

**Definitions of Properties**
性能的定义

**Density**: It is defined as mass per unit volume. It is expressed as $kg/m^3$.
密度：是指单位体积的质量。单位为 $kg/m^3$。

**Specific gravity**: It is the ratio of density of a material to density of water.
比重：是指材料密度与水密度之比。

**Porosity**: The term porosity is used to indicate the degree by which the volume of a material is occupied by pores.
孔隙率：是指材料中孔隙所占的体积比。

**Strength**: Strength of a material has been defined as its ability to resist the action of an external force without breaking.
强度：是指材料在无破坏条件下抵抗外界荷载作用的能力。

**Elasticity**: It is the property of a material which enables it to regain its original shape and size after the removal of external load.
弹性：是指材料在撤去外界荷载作用下能回复到原来的形状和尺寸的能力。

**Plasticity**: It is the property of the material which enables the formation of permanent deformation.
塑性：是指材料永久变形的能力。

**Hardness**: It is the property of the material which enables it to resist abrasion, indentation, machining and scratching.
硬度：是指材料所具备的抗磨损、刻蚀、切削等的能力。

**Ductility**: It is the property of a material which enables it to be drawn out or elongated to an appreciable extent before rupture occurs.

延性：是指材料所具备的在断裂前有一定程度的拉长及伸长能力。

***Brittleness***: It is the property of a material, which is opposite to ductility. Material, having very little property of deformation, either elastic or plastic is called Brittle.

脆性：是与延性相反的性质，指材料不具备变形的能力。

***Creep***: It is the property of the material which enables it under constant load to deform slowly but progressively over a certain period.

徐变：是指材料在长期荷载作用下不断缓慢变形的特性。

***Stiffness***: It is the property of a material which enables it to resist deformation.

刚度：是指材料抵抗变形的能力。

***Fatigue***: The term fatigue is generally referred to as the effect of cyclically repeated stress. A material has a tendency to fail at lesser stress level when subjected to repeated loading.

疲劳：是指在往复荷载作用下的响应；材料即使在承受小的重复荷载作用下也会发生破坏的特性。

***Impact strength***: The impact strength of a material is the quantity of work required to cause its failure per its unit volume. It thus indicates the toughness of a material.

冲击强度：是指单位体积的材料发生破坏时所需要对其做的功；表明材料的韧性。

***Toughness***: It is the property of a material which enables it to be twisted, bent or stretched under a high stress before rupture.

韧性：是指材料在断裂前承受扭转、弯曲及拉伸的性能。

***Thermal Conductivity***: It is the property of a material which allows conduction of heat through its body. It is defined as the amount of heat in kilocalories (or kilojoule) that will flow through unit area of the material with unit thickness in unit time when difference of temperature on its faces is also unity.

热传导：是指材料允许热量通过本身传递的物性。它被定义为，材料的表面在变化单位温度时，单位时间、单位面积、单位厚度上通过的热量，并以千卡(或千焦)为单位计量。

***Corrosion Resistance***: It is the property of a material to withstand the action of acids, alkalis gases etc., which tend to corrode (or oxidize).

耐腐蚀性：是指材料抵抗酸、碱等腐蚀作用(一般为氧化)的性能。

## 3.1　Traditional Materials　传统建筑材料

The building materials such as stones, bricks and timber are called traditional building materials since these materials are used from the early ages of civilization. Cement, cement-mortar and cement-concrete are also grouped under traditional materials although they are of recent origin compared to stones and bricks. The properties and uses of some of the most commonly used traditional building materials have been discussed here.

石料、砖及木材等材料在人类文明早期即被应用，故均属传统建筑材料。水泥、砂浆及混凝土等也被归为此类，尽管它们比砖、石等材料的应用较晚一些。一些最常用的传统建筑材料的性能如下所述。

## 1. Early construction materials  早期建筑材料

### 1) Stones  石材

Stones (as shown in Figure 3.1) are obtained from rocks. A rock represents a definite portion of earth's surface. Process of taking out stones from natural rock beds is known as quarrying. The term quarry is used to indicate the exposed surface of natural rocks. Stones, thus obtained, are used for various engineering purposes.

石材(如图 3.1 所示)来源于岩石。岩石能够表明地表土的组分。从天然岩层状态下取石料的过程称为"采掘"。采掘使岩层表面暴露，获得的石料用于多种工程的施工。

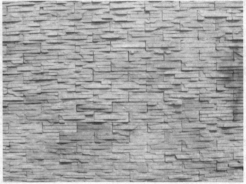

Figure 3.1  Stones
图 3.1  石材

Uses of Stones:
石材的用途：

① Stones are used as a construction material for foundations, walls, columns and lintels.

② Stones are used for face-work of buildings to give good appearance.

③ Thin stone slabs are used as roofing and flooring material.

④ Crushed stones are used for production of cement concrete.

⑤ Crushed stones are also used as ballast for railway track.

⑥ Aggregate of stone is used as a road metal.

①可用于基础、墙、柱、过梁的建筑材料；②可用于建筑外墙的饰面做法，使其美观；③薄的石板可用作楼屋面板；④破碎的石料可作为生产水泥的原料；⑤粗料石可用于铁轨的压载材料；⑥料石可作为路基材料。

### 2) Bricks  砖

Bricks (as shown in Figure 3.2) are obtained by molding clay in the rectangular blocks of uniform size and then by drying and burning these blocks. Bricks are very popular as they are easily available, economical, strong, durable and reliable. They are also reasonably heat and sound proof. Thus, at places where stones are not easily available, if there is plenty of clay, bricks replace stones.

砖(如图 3.2 所示)是通过在统一尺寸的矩形块中浇注泥土并烧结而成；砖因取材容易、经济性好、强度大、耐久性及可靠性而被广泛应用。砖是天然的隔热、隔音材料。在石料缺乏的地区，如果当地拥有大量的黏土资源，可以用砖来代替石料。

Figure 3.2　Bricks

图 3.2　砖

Uses of Bricks:

砖的用途：

① Bricks are extensively used as a leading material of construction.

② A fire brick (refractory brick) is used for lining the interiors of ovens, chimneys and furnaces.

③ Broken bricks are used as a ballast material for railway tracks, and also as a road metal.

④ Bricks are extensively used for construction of load-bearing walls and partition-walls.

⑤ Bricks are also used for face-work when artistic effect is required.

①砖是目前主要的建筑材料之一；②耐火砖可用于烤炉、烟囱、熔炉的内衬施工；③破碎的砖可用于铁路路基的填压材料；④砖广泛用于承重墙、隔墙的施工；⑤也可用于建筑需要的饰面做法。

3) Timber　木材

The word timber (as shown in Figure 3.3) is derived from Timbrian which means to build. Timber thus denotes wood which is suitable for building construction, carpeting or other engineering purposes. When it forms part of a living tree it is called "standing timber", when felled it is known as "rough timber" and when cut into different forms such as planks, beams etc., it is known as "converted timber".

木材(如图 3.3 所示)的命名来源于 Timbrian，意为建造。因此木材十分适合于建筑施工、地面铺装及其他的施工目的。当其还属于有生命的树木的一部分时被称为"建筑用材"，而若已砍伐下来时，则称为"原木"，而当其已被加工成不同的梁、柱部分时，称为"木构件"。

Figure 3.3  Timber and Products of Timber
图 3.3  木材和木制品

Uses of Timber:
木材的用途：

① For making doors, windows and ventilators.
② As flooring and roofing material.
③ For making furniture.
④ In making agricultural implements.
⑤ In the manufacture of sport goods, musical instruments etc.
⑥ In making railway coaches, wagons, buses, boats etc.
⑦ In making sleepers for railways, fencing poles, electric poles, footways, bridge floors, temporary bridges etc.
⑧ Soft wood is used for manufacture of paper, card-boards, walls paper etc.

①用作门、窗及通风设施；②用作楼、屋面板部分；③制作家具；④制作农业设备；⑤用于制作运动器械、乐器等；⑥用于铁路的车厢、汽车及船等部件；⑦用于铁路枕木、围墙及电线杆、步行街、桥面板及简易桥梁等；⑧软木用于制作纸、纸板、墙纸等。

4) Mortar
砂浆

The term mortar (as shown in Figure 3.4) is used to indicate a paste prepared by adding required quantity of water to a mixture of binding material and sand. The most common binding material used in the preparation of mortar is cement. Such a mortar is called cement-mortar.

砂浆(如图 3.4 所示)是通过向胶凝材料和砂的混合物中加入定量的水所获得的黏结材料。准备砂浆时最常用的胶凝材料就是水泥，这种砂浆称为水泥砂浆。

Figure 3.4 Mortar

图 3.4 砂浆

Uses of mortar:

砂浆的用途：

① As a binding material for brick and stone masonry constructions.

② To carry out pointing and plastering work on exposed surfaces of masonry.

③ To provide fine finish to concrete works.

④ For decorative finish to masonry walls and roof slabs.

⑤ As a filler material in ferro-cement works.

①砂浆可用作黏结砖或石块的黏结材料；②可用于砌体结构的外墙抹灰；③为混凝土提供细骨料；④用于砖墙或混凝土板的饰面工程；⑤可用于水泥施工中的填充料。

2. Modern construction materials  现代建筑材料

1) Cement concrete  混凝土

Cement concrete is an artificial product obtained by hardening of the mixture of cement, sand, gravel and water in pre-determined proportions. The hardening of cement-concrete is due to chemical reaction between cement and water and the process is also called "setting". Mixing cement concrete is shown in Figure 3.5.

混凝土是由水泥、砂子、石料及水按一定的比例混合经结硬后形成的一种人工建筑材料。混凝土的硬化主要是由水泥与水发生化学反应来完成，这一过程也称作"凝结"。混凝土的搅拌如图 3.5 所示。

Figure 3.5 Mixing Cement Concrete

图 3.5 混凝土的搅拌

Properties of cement-concrete:

混凝土的性能：

Properties of concrete are different when it is in plastic stage (green concrete) and hardened stage.

混凝土的性能在流动状态与硬化状态下截然不同。

Properties of green concrete:

新制备的混凝土性能：

① *Workability*: The workability of a freshly prepared concrete is the ease with which it can be mixed, placed, compacted and finished without affecting homogeneity.

和易性：新制备的混凝土的和易性是指其非常容易进行拌和、浇注、压实及施工而不会影响其性质。

② *Segregation:* A good concrete is a homogeneous mixture of cement, sand, coarse aggregates and water. Segregation during mixing, placing and compacting affects strength, durability and water tightness.

离析性：制备性能好的混凝土是水泥、水、砂子及石子形成了各向同性材料。在拌和、浇注及振捣过程中的离析会影响混凝土的强度、耐久性及水密性。

③ *Bleeding:* Bleeding is the separation of water cement slurry from coarse and fine aggregates either due to excessive quantity of water in fresh concrete or due to excessive vibration during compaction process. Bleeding of concrete causes formation of pores and reduces compressive strength.

裂隙性：泥浆与粗细骨料分离，主要是在制备混凝土过程中加入了过量的水或在施工过程中的过度振捣。这种情况会使得混凝土呈现多孔性，影响混凝土的强度。

④ *Harshness:* Harshness of concrete is due to use of poorly graded aggregates in concrete mix. A harsh concrete is porous and difficult to get smooth surface finish.

粗糙性：混凝土的粗糙性是混凝土在制备过程中由于骨料级配不良造成的。这种混凝土是多孔的且表面很难施工平整。

Properties of hardened concrete:

硬化后的混凝土性能：

① *Strength:* Hardened concrete should have high compressive strength. Compressive strength is the most important parameter considered in the design of concrete structures.

强度：结硬后的混凝土具有很高的抗压强度，抗压强度是混凝土结构设计的主要参数。

② *Durability:* Concrete should be capable of withstanding the adverse effects of weathering agents such as wind and water. In addition, it should resist temperature variation, moisture variation, freezing and thawing.

耐久性：混凝土本身能够抵抗外界天气的变化，如风和水的作用。能够承受温度变化、潮湿、冻融作用。

③ *Impermeability:* Hardened concrete should be impermeable or watertight. This property of concrete is considered in the construction of water tanks and bins.

密闭性：硬化后的混凝土是不亲水的，这一性质是混凝土制备、施工过程中需要考虑的。

④ ***Resistance to wear and tear:*** Hardened concrete should be capable of withstanding abrasive action during its usage. When concrete is used as a flooring material, or for construction of cement concrete roads, it should withstand abrasive action.

耐磨性：硬化的混凝土在使用过程中应该具备一定的耐磨性。若混凝土应用于楼面板结构中、水泥路面结构中，应能承受一定的磨损作用。

Uses of Cement Concrete:

混凝土的用途：

① It is used for making Reinforced Cement Concrete (R.C.C.) and Prestressed Cement Concrete (P.S.C.).

可用于钢筋混凝土结构和预应力混凝土结构。

② It is used for mass concrete works such as dams and bridges.

可用于大体积混凝土施工，如大坝、桥梁结构。

③ It is used for making electric poles, railway sleepers and high rise towers.

可用于电线杆、铁路轨木和高塔结构。

④ It is used for the construction of silos and bunkers.

可用于筒仓、煤仓等特种结构。

⑤ It is used for the construction of water tanks and under water construction.

可用于水箱及水下工程的施工。

⑥ It is used for the construction of road pavements and air port pavements.

可用于路面及机场跑道的施工。

⑦ It is used for construction of arches and ornamental structures.

可用于拱结构及装饰性结构的施工。

⑧ It is one of the best universally accepted construction material and used in various civil engineering works.

混凝土是受到最普遍认同的施工材料之一，被用于各种土木工程结构中。

2) Reinforced Cement Concrete (R.C.C.) 钢筋混凝土

Plain concrete can withstand very high compressive loads, but it is very weak in resisting tensile loads. Steel, on the other hand can resist very high tensile force. Hence, steel is embedded in concrete whenever tensile stresses are expected in concrete. Such a concrete is called "Reinforced Cement Concrete" (as shown in Figure 3.6).

素混凝土可抵抗很高的抗压荷载，但其抗拉强度很低。另一方面，钢筋可承受较高的拉应力。因此，钢筋被置于混凝土结构中可能出现拉应力的位置，使其承受拉应力。这种混凝土被称为"钢筋混凝土"（如图 3.6 所示）。

Figure 3.6 Reinforced Cement Concrete
图 3.6 钢筋混凝土

Uses of Reinforced Cement Concrete:
钢筋混凝土的用途：

① R.C.C. is used for construction of structural elements such as beams, columns and slabs.
可用于梁、柱及板的施工。

② R.C.C. is used for the construction of water tanks, storage bins, bunkers, tall chimneys, towers etc.
可用于水塔、储罐、筒仓、高烟囱及塔架等。

③ R.C.C. is used in making raft foundations and pile foundations.
可用于筏基及桩基施工。

④ R.C.C. is used in the construction of bridges, marine structures, aqueducts, high rise buildings and many other civil engineering works.
可用于桥梁、海岸工程的施工、高层建筑及其他土木工程等。

3) Ferro-Cement　钢纤维混凝土

Ferro cement (as shown in Figure 3.7) implies cement mortar of cement sand ratio 1:2 or 1:3 reinforced with multiple layers of steel fibres. The reinforcement is usually made of wire mesh which is usually made of 0.8 to 1.0mm diameter steel wires at 5mm to 50mm spacing. The skeletal steel may be placed 300 mm apart to serve as a spacer rod to the mesh.

钢纤维混凝土(如图 3.7 所示)指在水灰比为 1：2 或 1：3 的混凝土中加入多层钢纤维，通常是由直径为 0.8～1.0mm、间距为 5～50mm 的钢丝制成钢丝网。钢筋骨架间距 300mm 设置，可作为钢丝网的支垫。

Ferro-cement is suitable for low-cost roofing, pre-cast units, man-hole covers etc. It can be used for the construction of domes, vaults, shells, grid surfaces and folded plates. It is a good substitute for timber. It can be used for making furniture, doors and window frames, shutters and partitions. It can also be used for making water tanks, boats and silos.

钢纤维混凝土适合于低造价的屋盖结构、预制构件、检修孔等工程，可应用于穹顶、拱顶、壳结构及褶皱板等构件施工，是很好的木材替代材料。可用于家具、门窗及隔断等部位；也可制作水罐、船只及筒仓结构。

Figure 3.7　Ferro Cement
图 3.7　钢纤维混凝土

## 3.2　Composite Materials　复合材料

Composites are combinations of two more separate materials on a microscopic level, in a controlled manner to give desired properties. The properties of a composite will be different from those of the constituents in isolation. When two materials are combined together to form a composite, one of the materials will be in "reinforcing phase" and the other material will be in "matrix phase". Typically, reinforcing material in the form of fibers, sheets or particles are strong with low densities while the matrix is usually a ductile or tough material, e.g. glass - reinforcing material, polyester-matrix material.

Glass + Polyester=GRP (Glass Fibre Reinforced Plastic)

复合材料是在微观层次上用两种不同的材料、采用特定的方式来获得预期效果而形成的一种材料。复合材料有别于其他单一组分的材料。若两种不同的材料组合在一起，其中一种材料为增强相，而另一种材料为基体。通常，增强相是纤维材料、片状或颗粒状等强度高、密度低的材料，而基体材料往往是有延展性的或坚硬的材料，如玻璃纤维、聚酯材料等。

玻璃+聚酯 = GRP(玻璃纤维增强塑料)

### 1. Classification　分类

Composites can be broadly classified in to two groups. They are Natural composites and Man-made composites.

复合材料大体上可分为两类，一是自然生成的复合材料；二是人造复合材料。

Several natural materials can be grouped under natural composites, e.g. bone, wood etc.; man-made composites are produced by combining two or more materials in definite proportions under controlled conditions.

自然形成的复合材料种类很多，如骨骼，木材等；人造复合材料是通过使两种及以上的材料按一定的比例在一定的条件下形成的新材料。

## 2. Properties of Composites　复合材料的性能

① Composites possess excellent strength and stiffness.
② They are very light materials.
③ They possess high resistance to corrosion, chemicals and other weathering agents.
④ They can be molded to any shape and size with required mechanical properties in different directions.

①复合材料具有优越的强度及刚度性能；②自重轻；③抗腐蚀性强；④可根据在不同方向下的力学性能制造成不同的形状和尺寸。

Disadvantages of composites:
复合材料的缺点：
① High production cost.
② Difficult to repair.
③ Susceptible to damage.

造价高；难以修复；易损坏。

## 3. Uses of Composite Materials　在土木工程领域中的应用

① Extensively used in space technology and production of commercial airplanes.
② Used in the production of sport goods.
③ Used for general industrial and engineering structures.
④ Used in high speed and fuel efficient transport vehicles (The harmony express is shown in Figure 3.8).

①广泛应用于空间技术及商用飞机结构中；②用于运动器械上；③用于一般的工业及民用建筑中；④用于高速、高效的交通工具中(如图 3.8 所示为和谐号动车)。

Figure 3.8　The Harmony Express
图 3.8　和谐号动车

Composites are extensively used in the field of Civil Engineering. Ferro-cement is a good example for composite. R.C.C. and P.S.C. (Pre-stressed Cement Concrete) are composites that are widely used for structural components. Even concrete can be considered as a composite.

Block boards, Batten boards, and Chip boards which are composites are used in light construction works such as doors, windows, furniture and cabinets. Asbestos Cement Sheets are used as roofing material. Reinforced glass is used for sky-lights and door and window paneling.

复合材料广泛应用于土木工程领域。广泛应用于结构构件中的纤维混凝土即为一例。而混凝土本身也为一种组合材料。芯块、胶合板、缀板、芯板等均属复合材料并被广泛应用轻质构件，如门、窗、家具及橱柜中。防火棉用于屋盖结构中，强化玻璃应用于幕墙结构中等。

## 3.3　Smart Materials　智能材料

They are the materials which have the capability to respond to changes in their condition or the environment to which they are exposed, in a useful and usually repetitive manner. They are called by other names such as, intelligent materials, active materials and adoptive materials.

智能材料具备可适应所暴露或接触的天气、环境等变化的性能。又被称为活性材料等。

The devises that are made using smart materials are called "Smart Devices". Similarly the systems and structures that have incorporated smart materials are called "Smart Systems" (as shown in Figure 3.9) and "Smart Structures". In other words the complexity increases from smart materials to smart structures.

用智能材料制作的设备称为"智能设备"。相应地，采用智能材料的系统和结构分别被称为"智能系统"(如图3.9所示)和"智能结构"。换言之，目前已从智能材料阶段发展到更为复杂的智能结构。

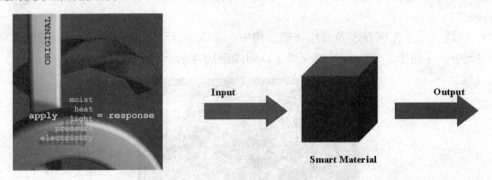

Figure 3.9　Smart System
图3.9　智能系统

1. Stimulus Response System (as shown in Figure 3.10)　刺激-反应系统(如图3.10所示)

A smart material or an active material gives an unique output for a well defined input. The input may be in the form of mechanical stress/strain, electrical/magnetic field or changes in temperature. Based on input and output, the smart materials are classified as following.

智能材料对于定义良好的"输入"会提供一个独特的"输出"。输入可能是以应力或应变、电或磁以及温度的变化等形式给出。基于"输入"及"输出"的内容，智能材料可分为如下几类。

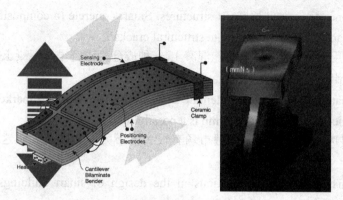

Figure 3.10  Stimulus Response System
图 3.10  刺激-反应系统

1) Shape Memory Alloys (SMAs)  形状记忆合金

They are the smart materials which have the ability to return to some previously defined shape or size when subjected to appropriate thermal changes, e.g. Titanium-Nickel Alloys.

这种智能材料在适合的热交换条件下，可自动地恢复为原有的形状或尺寸，如钛-镍合金。

2) Magnetostrictive Materials  磁致伸缩材料

They are the smart materials which have the ability to undergo deformation when subjected to magnetic field, e.g. Terfenol-D (Alloy of Iron and Terbium).

这种智能材料当置于磁场中时，可产生一定的变形，如稀土超磁致伸缩材料(铁和稀土合金)。

3) Piezoelectric Materials  压电材料

These are the materials which have capability to produce a voltage when surface strain is introduced. Conversely, the material undergo deformation (stress) when an electric field is applied across it.

这种智能材料当表面产生一定的应变时会产生一定的电压；反之，若表面存在电势场时，该材料也可产生变形。

4) Electrorheological Fluids  电流变材料

They are the colloidal suspensions that undergo changes in viscosity when subjected to an electric field. Such fluids are highly sensitive and respond instantaneously to any change in the applied electric field.

电流变材料属于胶质悬浮材料，当处于电势场中时表现出一定的黏性。电势场中任何微小的电势变化都会引起其灵敏及快速的反应。

2. Applications  智能材料的用途

(1) Smart materials are used in aircrafts and spacecrafts to control vibrations and excessive deflections.

主要用于航空、航天器中来控制振动及挠度过大。

(2) Smart concrete is used in smart structures. Smart concrete (a composite of carbon fibres and concrete) is capable of sensing minute structural cracks/flaws.

智能混凝土用于智能结构中。智能混凝土(一种成分为碳纤维与混凝土的复合材料)能够监测微小裂缝的出现。

(3) Smart materials have good potential to be used in health care markets. Active control drug delivery devices such as Insulin Pump is a possibility.

将智能材料应用于医疗卫生保健市场有很大的潜力，如胰岛素泵这种主动控制给药装置。

(4) Smart materials have applications in the design of smart buildings and state of art vehicles. Smart materials are used for vibration control, noise mitigation, safety and performance improvement.

智能材料应用于智能建筑和一流的交通工具的设计中。智能材料可用于振动控制、噪声抑制、安全性和性能的提高等方面。

# New Words and Expressions 生词和短语

1. brick     n. 砖，砖块；砖形物
2. timber    n. 木材；木料
3. cement    n. 水泥；接合剂
4. steel     n. 钢铁；钢制品；坚固
5. alternate building material     交替建筑材料
6. composite material     复合材料
7. smart material     智能材料
8. density    n. 密度
9. porosity   n. 有孔性，多孔性
10. specimen    n. 样品，样本；标本
11. deformation   n. 变形
12. abrasion    n. 磨损；磨耗；擦伤
13. ductility   n. 延展性；柔软性
14. rupture   n. 破裂；决裂；疝气
15. brittleness   n. [材] 脆性，[材] 脆度；脆弱性
16. creep    n. 爬行；蔓延；蠕变
17. stiffness   n. 刚度；硬度
18. fatigue   n. 疲劳，疲乏
19. toughness   n. [力] 韧性；强健；有黏性
20. corrosion resistance     耐蚀性；抗腐蚀性
21. cement-mortar     水泥砂浆
22. cement-concrete     水泥混凝土

23. quarry    n. 采石场
24. lintel    n. [建] 过梁
25. moulding clay    浇铸泥土
26. load-bearing wall    承重墙
27. partition-wall    隔墙
28. mortar    n. 砂浆
29. gravel    n. 碎石；沙砾
30. workability    n. 可使用性；施工性能；可加工性
31. segregation    n. 隔离，分离
32. harshness    n. 严肃；刺耳；粗糙的事物
33. durability    n. 耐久性；坚固；耐用年限
34. impermeability    n. 不渗透性；不透过性
35. Stimulus Response System    刺激-反应系统
36. magnetostrictive    adj. [物] 磁致伸缩的
37. piezoelectric    adj. [电] 压电的
38. Electrorheological Fluid    电流变液体

# Exercises　练习

Ⅰ. Write a T in front of a statement if it is true according to the text and write an F if it is false.

1. Specific gravity is the ratio of density of a material to density of water.
2. Thermal Conductivity is defined as the amount of heat in unit calories that will flow through unit area of the material with unit thickness in unit time when difference of temperature on its faces is also unity.
3. Broken bricks are used as a ballast material for railway tracks, and also as a road metal.
4. Cement Concrete can be used for making electric poles, railway sleepers and high rise towers.
5. Composite Materials possess low resistance to corrosion, chemicals and other weathering agents.
6. Magnetostrictive Materials are the smart materials which have the ability to undergo deformation when subjected to magnetic field.

Ⅱ. Complete the following sentences.

1. Engineering Materials or Building Materials include _____, _____, _____, _____ and _____.
2. Some of the most important properties of building materials are _____, _____, _____, _____, _____, _____ and _____.

3. The traditional building materials include _____, _____ and _____.

4. The most common binding material used in the preparation of mortar is _____.

5. Properties of Green Concrete are _____, _____, _____, _____.

6. Disadvantages of composites Materials include _____, _____, _____.

7. Smart Materials are also called _____ or _____ or _____.

Ⅲ. Translate the following sentences into Chinese.

1. The materials used in the construction of Engineering Structures such as buildings, bridges and roads are called Engineering Materials or Building Materials.

2. Stiffness is the property of a material which enables it to resist deformation.

3. Bricks are obtained by moulding clay in the rectangular blocks of uniform size and then by drying and burning these blocks.

4. The term mortar is used to indicate a paste prepared by adding required quantity of water to a mixture of binding material and sand.

5. Cement Concrete is one of the best universally accepted construction material and used in various civil Engineering works.

6. A smart material or an active material gives an unique output for a well defined input.

Ⅳ. Translate the following sentences into English.

1. 一个工程师熟悉工程材料的性能是必要的。

2. 建筑材料，如石头、砖块和木材被称为传统的建筑材料。

3. 混凝土在塑性阶段和固化阶段的性能是不同的。

4. 复合材料大致可分为两组，分别是天然复合材料和人造材料。

5. 复合材料可以在不同的方向按照机械性能的要求以任意形状和大小尺寸加工成型。

6. 智能材料应用于智能建筑和一流的交通工具的设计中。

## Answers 答案

Ⅰ.

1. T  2. F  3. T  4. T  5. F  6. T

Ⅱ.

1. brick, timber, cement, steel, plastic

2. physical, mechanical, thermal, chemical, optical, acoustical, physiochemical

3. stones, bricks, timber

4. cement

5. workability, segregation, bleeding, harshness

6. high production cost, difficult to repair, susceptible to damage

7. Intelligent materials, Active materials ,Adoptive materials

Ⅲ.

1. 用于像建筑物、桥梁和道路等工程结构建设方面的材料叫作工程材料或建筑材料。

2. 硬度是材料抵抗变形的性能。

3. 砖是通过在统一尺寸的矩形块中浇注泥土并烧结而成。

4. 术语砂浆是用来表示把所需要的用水量加入到黏结材料和砂的混合物中形成的糊状物。

5. 水泥混凝土是最被普遍接受的建筑材料之一，可以用于各种土建工程。

6. 智能材料或活性物质根据确定的输入而产生一个特定的输出。

Ⅳ.

1. It is necessary for an engineer to be conversant with the properties of engineering materials.

2. The building materials such as stones, bricks and timber are called traditional building materials.

3. Properties of concrete are different when it is in plastic stage and hardened stage.

4. Composites materials can be broadly classified in to two groups. They are Natural composites materials and Man-made composites materials.

5. Composites materials can be molded to any shape and size with required mechanical properties in different directions.

6. Smart materials have applications in the design of smart buildings and state of art vehicles.

# Chapter 4　Loads
# 第 4 章　荷载

The main purpose of a structure is to transfer load from one point to another: bridge deck to pier; slab to beam; beam to girder; girder to column; column to foundation; foundation to soil.

There can also be secondary loads such as thermal (in restrained structures), differential settlement of foundations, P-Delta effects (additional moment caused by the product of the vertical force and the lateral displacement caused by lateral load in a high rise building).

结构的主要作用就是传递荷载，如从桥面结构到桥墩、板到梁、梁到桁、梁到柱、柱到基础、基础到地基的传递。

还有其他形式的附加荷载，如热(在超静定结构中)，基础的不均匀沉降，重力二阶效应(指在高层建筑中，竖向荷载在水平荷载所产生的位移上所形成的附加弯矩)等。

Loads are generally subdivided into two categories:
荷载可分为两类：
Vertical loads or gravity load:
① dead load (DL).
② live load (LL).
③ snow load.
竖向荷载或重力荷载：包括静荷载、活荷载、雪荷载。
Lateral loads which act horizontally on the structure:
① wind load (WL).
② earthquake load (EL).
③ hydrostatic and earth loads.
水平荷载：包括风荷载、地震荷载、土压力。

This distinction is helpful not only to compute a structure's load, but also to assign different factor of safety to each one. For a detailed coverage of loads, refer to the *Universal Building Code* (UBC) (UBC 1995).

这种分类可不仅有利于荷载的计算，也可对每种荷载进行安全系数的确定。其他类型的荷载可参阅通用建筑规范(UBC 1995)。

## 4.1　Vertical Loads　竖向荷载

For closely spaced identical loads (such as joist loads), it is customary to treat them as a uniformly distributed load rather than as discrete loads, as shown in Table. 4.1, Table 4.2,

Table 4.3。

对于分布较集中的荷载(如托梁的荷载)，习惯上将其视为均匀分布的荷载，而非离散荷载。如表 4.1～表 4.3 所示。

Table 4.1　Vertical Load

表 4.1　竖向荷载

| Material<br>材料 | Unit (lb/ft$^2$)<br>单位(磅/英尺$^2$) |
|---|---|
| Ceilings<br>天花板 | |
| Channel suspended system<br>通道的悬浮系统 | 1 |
| Acoustical fiber tile<br>声学纤维瓦 | 1 |
| Roofs<br>楼面结构 | |
| Copper or tin<br>铜或锡 | 1～5 |
| 5 ply felt and gravel<br>5 层毡及卵石层 | 6 |
| Shingles asphalt<br>屋顶板 | 3 |
| Clay tiles<br>黏土瓦 | 9～14 |
| Sheathing wood<br>木盖层 | 3 |
| Insulation 1 in poured in place<br>隔热层 | 2 |
| Partitions<br>构造部分 | |
| Clay tile 3 in<br>3 英寸黏土瓦 | 17 |
| Clay tile 10 in<br>10 英寸黏土瓦 | 40 |
| Gypsum block 5 in<br>5 英寸石膏 | 14 |
| Wood studs 2×4 (12-16 in)<br>木栓 2×4 | 2 |
| Plaster 1 in cement<br>1 英寸水泥砂浆 | 10 |
| Plaster 1 in gypsum<br>1 英寸石膏 | 5 |

Continued 续表

| | |
|---|---|
| Walls 墙面构件 | |
| Bricks 4 in<br>4 英寸砖 | 40 |
| Bricks 12 in<br>12 英寸砖 | 120 |
| Hollow concrete block (heavy aggregate)<br>混凝土空心砌块(重骨料) | |
| 4 in<br>4 英寸 | 30 |
| 8 in<br>8 英寸 | 55 |
| 12 in<br>12 英寸 | 80 |
| Hollow concrete block (light aggregate)<br>混凝土空心砌块(轻骨料) | |
| 4 in<br>4 英寸 | 21 |
| 8 in<br>8 英寸 | 38 |
| 12 in<br>12 英寸 | 55 |

Table 4.2  Weights of Building Materials
表 4.2  结构自重

| Material<br>材料 | Unit l(b/ft$^2$)<br>单位(磅/英尺$^2$) |
|---|---|
| Timber<br>木材 | 40~50 |
| Steel<br>钢材 | 50~80 |
| Reinforced concrete<br>钢筋混凝土 | 100~150 |

As seen in Table 4.3, during design of column section, 740kips should be reduced to 410kips.

The total dead load: DL=10×60=600kips.Thus overall reduced by 25%.

如表 4.3 所示，柱底截面设计时，需将 740kips 折减至 410kips。

总恒载：DL=10×60=600kips，则总的折减为 25%。

Table 4.3　Average Gross Dead Load in Buildings

表 4.3　建筑物平均总恒载

| Floor<br>楼板 | Roof<br>屋面 | 10 | 9 | 8 | 7 | 6 | 5 | 4 | 3 | 2 | Total<br>汇总 |
|---|---|---|---|---|---|---|---|---|---|---|---|
| Cumulative $R$ (%)<br>累积 $R$ (%) | 8.48 | 16.96 | 25.44 | 33.92 | 42.4 | 51.32 | 59.8 | 60 | 60 | 60 | |
| Cumulative LL<br>累积 LL | 20 | 80 | 80 | 80 | 80 | 80 | 80 | 80 | 80 | 80 | 740 |
| Cumulative $R$×LL<br>累积 $R$×LL | 18.3 | 66.4 | 59.6 | 52.9 | 46.08 | 38.9 | 32.2 | 32 | 32 | 32 | 410 |

**Snow load**　雪载

Roof snow load vary greatly depending on geographic location and elevation. They range from 20 to 45 psf, as shown in Figure 4.1.

屋面雪荷载主要取决于地理位置及海海拔高度。其取值为 20～45psf(注：1psf=1lb/ft$^2$)，如图 4.1 所示。

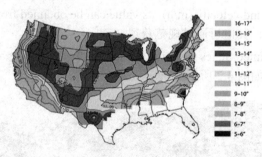

Figure 4.1　Snow Map of the United States

图 4.1　美国雪荷载分布图

Snow loads are always given on the projected length or area on a slope.

The steeper the roof, the lower the snow retention. For snow loads greater than 20 psf and roof pitches more than 20°, the snow load $P$ may be reduced by

雪荷载通常指是的一倾斜屋面上一定长度或面积上的雪自重。

屋面倾角越大，雪荷载越小。当雪荷载超过 20psf 及屋面倾角大于 20°时，雪荷载 $P$ 可按下式折减计算：

$$R = (\alpha - 20)\left(\frac{P}{40} - 0.5\right) \quad \text{(psf)} \tag{4.1}$$

## 4.2 Lateral Loads  水平荷载

### 1. Wind  风载

Wind load depends on: velocity of the wind, shape of the building, height, geographical location, texture of the building surface and stiffness of the structure. Wind loads are particularly significant on tall buildings.

结构所承受的风载主要取决于风速、建筑的形状、高度、地理位置、建筑表面形状及建筑结构的刚度。风荷载对结构的影响十分重要。

When a steady streamline airflow of velocity $V$ is completely stopped by a rigid body, the stagnation pressure (or velocity pressure) $q_s$ was derived by Bernouilli:

当一个以速度 $V$ 行进的稳定的气流完全被一刚性建筑所阻碍时，则风的压力可由伯努利方程获得：

$$q_s = \frac{1}{2}\rho V^2 \tag{4.2}$$

where the air mass density $\rho$ is the air weight divided by the acceleration of gravity $g=32.2\text{ft/sec}^2$. At sea level and a temperature of 15°C (59°F), the air weighs 0.0765 lb/ft³ and this would yield a pressure of

其中空气密度 $\rho$ 为空气容重除以重力加速度 $g=32.2\text{ft/s}^2$。在海平面及 15 摄氏度(59 华氏度)时，空气密度为 0.0765 lb/ft³，由此产生的压力为

$$q_s = \frac{1}{2}\rho V^2 \tag{4.2}$$

$V$ is the maximum wind velocity (m/h), its value can be obtained from code.

$V$ 为最大风速(m/h)，其值可通过规范获得。

Wind maps in the United States is shown in Figure 4.2.

美国风荷载分布图如图 4.2 所示。

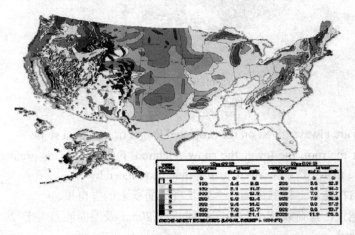

Figure 4.2  Wind Maps in the United States
图 4.2  美国风荷载分布图

The wind pressure can induce wind suction, as seen in Eqn.4.3.
风压在背风面可引起风吸力，如公式(4.3)所示。

During storms, wind velocities may reach values up to or greater than 150 miles per hour, which corresponds to dynamic pressure $q_s$ of about 60 psf (as high as the average vertical occupancy load in buildings). Wind pressure increases with height. Wind load will cause suction on the leeward sides. When it is in the horizontal plane and the temperature is 15℃, the wind pressure is

在暴风雨天气，风速可达每小时 150 英里及以上，其相应的动风压可达 60psf(与建筑物的平均垂直占有负荷相当)。风压随高度的增加而增加。风荷载在背风面产生吸力。在水平面并且气温为 15℃时，风压为

$$q_s = \frac{1}{2} \frac{(0.0765) \text{lb}/\text{ft}^3}{(32.2) \text{ft}/\text{sec}^2} \left( \frac{(5280) \text{ft}/\text{mile}}{(3600) \text{sec}/\text{hr}} V \right)^2 \quad (4.3)$$

or
或

$$q_s = 0.00256 V^2 \quad (4.4)$$

This magnitude must be modified to account for the shape and surroundings of the building. Thus, the design pressure $p$ (psf) is given by

考虑到建筑物的形状和周围环境，该值须进行修正。则设计风压可由下式计算：

$$P = C_e C_q I q_s \quad (4.5)$$

The pressure is assumed to be normal to all walls and roofs and $C_e$ Velocity Pressure Coefficient accounts for height, exposure and gust factor. It accounts for the fact that wind velocity increases with height and that dynamic character of the airflow (i.e. the wind pressure is not steady), as shown in Table 4.4.

假设风压沿墙及屋顶均匀分布。其中 $C_e$ 为风速系数，由高度、地面粗糙度及狂风系数确定。事实上，风速随高度及动力系数的增大而逐渐增加(即风压是不稳定的)，如表 4.4 所示。

Table 4.4  $C_e$ Coefficients for Wind Load

表 4.4  $C_e$ 风速系数

| $C_e$ | Exposure 地面粗糙度 | |
|---|---|---|
| 1.39~2.34 | D | Open, flat terrain facing large bodies of water<br>面向大片水域的空旷、平整的地带 |
| 1.06~2.19 | C | Flat open terrain, extending one-half mile or open from the site in any full quadrant<br>空旷、平整的面积在 1.5 英里以上的地区 |
| 0.62~1.80 | B | Terrain with buildings, forest, or surface irregularities 20 ft or more in height<br>存在房屋、森林，表面不规则且高度超过 20 米的地带 |

$C_q$ Pressure Coefficient is a shape factor which is given in Table 4.5 for gabled frames.

$C_q$ 压力系数是如表 4.5 所示的门式钢架的形状系数。

Table 4.5  Wind Pressure Coefficients $C_q$ (UBC 1995)

表 4.5  $C_q$ 风压系数

| | | Windward Side 迎风面 | Leeward Side 背风面 |
|---|---|---|---|
| Gabled Frames (V:H) 有山墙的框架($V:H$) | | | |
| Roof Slope 屋面倾角 | <9:12 | −0.7 | −0.7 |
| | 9:12 to 12:12<br>9：12～12：12 | 0.4 | −0.7 |
| | >12:12 | 0.7 | −0.7 |
| Walls 墙 | | 0.8 | −0.5 |
| Buildings (height<200 ft) 建筑结构($H$<200 英尺) | | | |
| Vertical Projections 垂直投影 | height<40 ft<br>$H$<40 英尺 | 1.3 | −1.3 |
| | height>40 ft<br>$H$>40 英尺 | 1.4 | −1.4 |
| Horizontal Projections 水平投影 | | −0.7 | −0.7 |

**Importance Factor**  重要性系数

(1) Essential Facilities: Hospitals; Fire and police stations; Tanks; Emergency vehicle shelters; standby power-generating equipment; Structures and equipment in government; Communication centers.

(2) Hazardous Facilities: Structures housing, supporting or containing sufficient quantities of toxic or explosive substances to be dangerous to the safety of the general public if released.

(3) Special occupancy structure: Covered structures whose primary occupancy is public assembly, capacity > 300 persons.

(1) 必要设施：如医院、消防局及警察局、水箱、急救中心、备用的能源设施、政府部门建筑物、新闻中心等；

(2) 危险设施：存放大量一旦释放会对公众造成危害的有毒的或放射性物质的建筑；

(3) 特殊用途的建筑：如容纳超过 300 人的公共避难场所等。

Approximate design wind pressure $P$ for ordinary wind force resisting building structures is shown in Figure 4.3.

一般建筑物的近似设计风压值如图 4.3 所示。

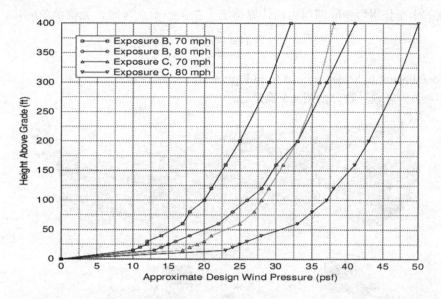

Figure 4.3　Approximate Design Wind Pressure *P* for Ordinary Wind Force Resisting Building Structures

图 4.3　一般建筑物的近似设计风压值

## 2. Seismic Load　地震荷载

Vibrations of a building is shown in Figure 4.4.

建筑物的结构振型如图 4.4 所示。

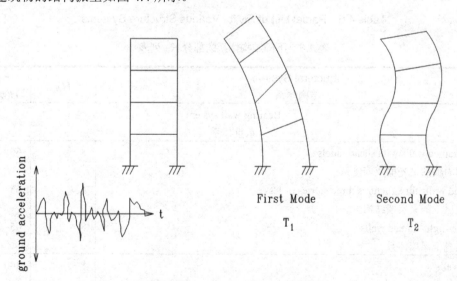

Figure 4.4　Vibrations of A Building

图 4.4　建筑物的结构振型

The horizontal force at each level is calculated as a portion of the base shear force. There is no whiplash effect when $T<0.7s$.

每层的水平作用可由底部剪力法计算得出。在 $T$ 小于 0.7s 时，无鞭梢效应。

$$V = \frac{ZJC}{R_W}W \tag{4.6}$$

$Z$: Zone Factor: to be determined from Figure 4.5.

其中：$Z$ 为区域系数，由图 4.5 确定。

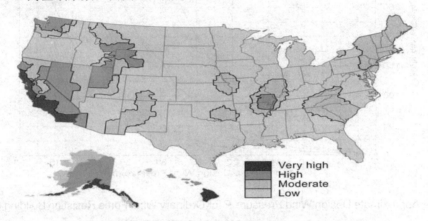

Figure 4.5 Seismic Zones of the United States(UBC 1995)

图 4.5 美国地震区域(UBC 1995)

Partial list of $R_W$ for various structure systems is shown in Table 4.6.

不同结构体系的部分 $R_W$ 列表如表 4.6 所示。

Table 4.6 Partial List of $R_W$ for Various Structure Systems

表 4.6 不同结构体系的部分 $R_W$ 列表

| Structural System 结构体系 | $R_W$ | $H$(ft) $H$(英尺) |
|---|---|---|
| Bearing wall system 承重体系 | | |
| Light-framed walls with shear panels 带有剪力板的采光框架墙体系 | 8 | 65 |
| Plywood walls for structures three stories or less 三层及以下的夹层墙结构 | 6 | 65 |
| All other light-framed walls 其他采光框架墙体系 | 8 | 65 |
| Shear walls 剪力墙体系 | 8 | 240 |
| Concrete 混凝土剪力墙 | 8 | 240 |
| Masonry 砌体剪力墙 | 8 | 160 |

Continued 续表

| Structural System<br>结构体系 | $R_W$ | $H$(ft)<br>$H$(英尺) |
|---|---|---|
| Building frame system (using trussing or shear walls)<br>含桁架或剪力墙的框架结构 | | |
| Steel eccentrically braced ductile frame<br>钢偏心支撑的延性框架 | 10 | 240 |
| Light-framed walls with shear panels<br>带有剪力板的采光框架墙体系 | 10 | 240 |
| Plywood walls for structures three stories or less<br>三层及以下的夹层墙结构 | 9 | 65 |
| All other light-framed walls<br>其他采光框架墙体系 | 7 | 65 |
| Shear walls<br>剪力墙体系 | 8 | 240 |
| Concrete<br>混凝土剪力墙 | 8 | 240 |
| Masonry<br>砌体剪力墙 | 8 | 160 |
| Concentrically braced frames<br>中心支撑的框架体系 | 8 | 160 |
| Steel<br>钢 | 8 | — |
| Concrete (only for zones 1 and 2)<br>混凝土 | 8 | 65 |
| Heavy timber<br>重木 | 8 | 65 |
| Moment-resisting frame system<br>抗弯框架体系 | | |
| Special moment-resisting frames (SMRF)<br>特殊的抗弯框架体系 | 12 | NL |
| Steel<br>钢 | 12 | NL |
| Concrete<br>混凝土 | 12 | NL |
| Concrete intermediate moment-resisting frames (IMRF) only for zones 1 and 2<br>只用于抗震Ⅰ类及Ⅱ类的混凝土抗弯框架体系 | 12 | NL |
| Ordinary moment-resisting frames (OMRF)<br>常规的抗弯框架体系 | 8 | — |
| Steel<br>钢 | 8 | 160 |
| Concrete (only for zone 1)<br>混凝土(只用于抗震Ⅰ类) | 6 | 160 |

| Structural System 结构体系 | $R_W$ | $H$(ft) $H$(英尺) |
|---|---|---|
| Dual systems (selected cases are for ductile rigid frames only) 双重抗侧体系 | | |
| Shear walls 剪力墙 | 12 | NL |
| Concrete with SMRF 带有特殊抗弯框架的混凝土剪力墙体系 | 12 | NL |
| Masonry with SMRF 带有特殊抗弯框架的砌体剪力墙体系 | 8 | 160 |
| Steel eccentrically braced ductile frame 钢偏心支撑的延性框架 | 6~12 | 160~NL |
| Concentrically braced frame 中心支撑的框架 | 12 | NL |
| Steel with steel SMRF 带有特殊抗弯框架的钢中心支撑体系 | 10 | NL |
| Steel with steel OMRF 带有常规抗弯框架的钢中心支撑体系 | 6 | 160 |
| Concrete with concrete SMRF (only for zones 1 and 2) 带有特殊混凝土抗弯框架的混凝土中心支撑体系(只用于抗震Ⅰ类及Ⅱ类) | 9 | — |

The total vertical load is
总的竖向荷载

$$W = 2 \times (200 + 0.5 \times 400) \times 20 = 16\,000 \text{ lb}$$

The total seismic base shear is
总的地震底部剪力

$$V = \frac{ZJC}{R_W} = \frac{0.3 \times 1.25 \times 2.75}{12} \approx 0.088W = 1375 \text{ lb}$$

Since $T<0.7$ sec there is no whiplash.
由于 $T<0.7$ 秒，不存在鞭梢效应。

The load on each floor is thus given by
因此每层的负载可由下式确定：

$$F_2 = \frac{1375 \times 24}{12 + 24} \approx 916.7 \text{lb}$$

$$F_1 = \frac{1375 \times 12}{12 + 24} \approx 458.3 \text{lb}$$

## 4.3 Other Loads 其他性质的荷载

### 1. Hydrostatic and Earth 静水压力及土压力

Structures below ground must resist lateral earth pressure.

位于地下的结构必须抵抗侧向土压力。

$$q = K\gamma h \tag{4.7}$$

where $\gamma$ is the density, $K = \dfrac{1-\sin\varphi}{1+\sin\varphi}$ is the pressure coefficient, $h$ is the height.

For sand and gravel $\gamma$=120lb/ft$^3$, and $\varphi \approx 30°$。

其中：$\gamma$为土的重度；$K$为土压力系数，$h$为埋深；

砂与石的等效容重：$\gamma$=120lb/ft$^3$；$\varphi$=30°。

If the structure is partially submerged, it must also resist hydrostatic pressure of water, as shown in Figure 4.6.

如果结构局部没入水下，则还需考虑地下水的静水压力，如图 4.6 所示。

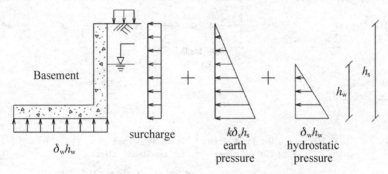

Figure 4.6 Earth and Hydrostatic Loads on Structures

图 4.6 结构上的土压力和静水压力

$$q = \gamma_w h \tag{4.8}$$

where $\gamma_w$ = 62.4lb/ft$^3$.

其中：$\gamma_w$=62.4lb/ft$^3$；

**Example 4-1:** Hydrostatic Load

例 4-1：静水压力

The basement of a building is 12 ft below ground. Ground water is located 9 ft below ground, thickness of concrete slab is required to exactly balance the hydrostatic uplift.

建筑的地下室埋深 12 英尺，地下水位于地表下 9 英尺处。试问，至少需用多厚的混凝土板才可确保其抗浮能力？

Solution: The hydrostatic pressure must be countered by the pressure caused by the weight of concrete.

分析：静水压力与混凝土自重是等值反向的。

Since $p=h$, we equate the two pressures and solve for $h$ the height of the concrete slab.

因 $p=h$，可利用二者相等的条件确定混凝土板厚。

$$h = \frac{(62.4)\text{lb}/\text{ft}^3}{(150)\text{lb}/\text{ft}^3}(3)\text{ft}(12)\text{in}/\text{ft} = 14.976\text{in} \approx 15.0\text{in}$$

## 2. Bridge Loads 桥梁荷载

For highway bridges, design loads are given by the AASHTO (Association of American

State Highway Transportation Officials). The HS-20 truck is used for the design of bridges on main highways, as shown in Figure 4.7. Either the design truck with specified axle loads and spacing must be used or the equivalent uniform load and concentrated load. This loading must be placed such that maximum stresses are produced.

对于高速桥梁，其设计荷载可由 AASHTO(美国公路协会)确定。主要高速公路的桥梁荷载的设计值依 HS-20 卡车荷载而定(如图 4.7 所示)。设计车辆荷载轴重及轮距应等效成均布荷载或集中荷载。该荷载应确保布置于能够产生最大的应力位置处。

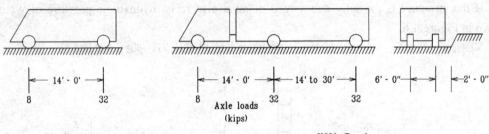

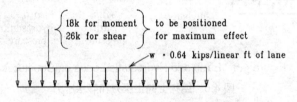

Figure 4.7　Truck Load

图 4.7　卡车荷载

# New Words and Expressions　生词和短语

1. deck　　　*n.* 甲板；层面 ；板面
2. pier　　　*n.* 码头，直码头；桥墩
3. slab　　　*n.* 厚板，平板；混凝土路面；厚片
4. beam　　　*n.* 横梁
5. girder　　*n.* [建] 大梁，纵梁
6. column　　*n.* 纵队，列；专栏；圆柱，柱形物
7. vertical load　　垂直荷载
8. gravity load　　重力荷载
9. lateral load　　横向荷载
10. hydrostatic and earth load　　静力和土荷载
11. joist load　　搁栅(或托梁)负荷
12. discrete load　　离散负载

13. roof pitch 屋面坡度；屋顶高跨比
14. stiffness     n. 僵硬；坚硬；刚度；硬度
15. velocity      n. [力] 速率；速度；周转率
16. stagnation pressure   [流] 滞止压力，[流] 驻点压力
17. dynamic pressure     动压力
18. magnitude     n. 大小；量级；[地震] 震级；重要；光度
19. coefficient   n. [数] 系数；率；协同因素
20. seismic load    地震载荷；地震力
21. storey       n. 楼层
22. ductile      adj. 柔软的；易教导的；易延展的
23. longitudinal direction    纵向；轴向
24. dimension    n. [数] 维；尺寸；次元；容积
25. thickness    n. 厚度；层；浓度
26. hydrostatic uplift     静水浮力
27. transmit     vt. 传输；传播；发射；传达

# Exercises 练习

Ⅰ．Write a T in front of a statement if it is true according to the text and write an F if it is false.

1. Hydrostatic and earth loads belongs to vertical load or gravity load.
2. It is customary to treat identical loads as a uniformly distributed loads rather than as discrete loads.
3. During storms, wind velocities may reach values up to or greater than 150 miles per hour, which corresponds to a dynamic pressure qs of about 60 psf.
4. Building codes allow certain reduction when certain loads are combined together.
5. Structures must be designed to resist a combination of loads.

Ⅱ．Complete the following sentences.

1. The main purpose of a structure is to transfer load from one point to another, such as _____.
2. Loads are generally subdivided into two categories:_____.
3. Lateral loads mainly include _____.
4. Covered structures whose primary occupancy is public assembly, capacity > _____persons.
5. The horizontal force at each level is calculated as _____.

Ⅲ．Translate the following sentences into Chinese.

1. There can also be secondary loads such as thermal, differential settlement of foundations,

P-Delta effects.

2. This distinction is helpful not only to compute a structure's load, but also to assign different factor of safety to each one.

3. For closely spaced identical loads, it is customary to treat them as a uniformly distributed load rather than as discrete loads.

4. It accounts for the fact that wind velocity increases with height and that dynamic character of the airflow.

5. When ground water is located 9 ft below ground, what thickness concrete slab is required to exactly balance the hydrostatic uplift?

Ⅳ. Translate the following sentences into English.

1. 一个结构的主要目的是从一点到另一点传递荷载。

2. 屋面雪荷载很大程度上取决于地理位置和海拔。

3. 本办公楼坐落在风速为每小时70英里、地震带4区的城市环境中。

4. 静水压力一定和混凝土重量引起的压力相反。

5. 当某些荷载组合在一起时，建筑法规允许对荷载进行一定的折减。

# Answers　答案

Ⅰ.

1. F　2. T　3. T　4. T　5. F

Ⅱ.

1. bridge deck to pier; slab to beam; beam to girder; girder to column; column to foundation; foundation to soil

2. vertical load or gravity load , lateral load

3. wind load (WL); earthquake load (EL) ; hydrostatic and earth loads

4. 300

5. a portion of the base shear force V

Ⅲ.

1. 也可以有二次负载，如热量、地基不均匀沉降、重力二阶效应。

2. 这种区别不仅有助于计算结构的负荷，而且还有助于指定每一个不同的安全系数。

3. 对于间隔很近的相同荷载，习惯上将其看作一个均匀分布的荷载而不是离散的荷载。

4. 它印证了这样的事实：风速随高度增加和气流的动态特性。

5. 当地下水位于地表以下9英尺时，需要多厚的混凝土板才能精确地平衡静水浮力？

Ⅳ.

1. The main purpose of a structure is to transfer load from one point to another.

2. Roof snow load vary greatly depending on geographic location and elevation.

3. This office building is located in an urban environment with a wind velocity of 70 mph and in seismic zone 4.

4. The hydrostatic pressure must be countered by the pressure caused by the weight of concrete.

5. Building codes allow certain reduction when certain loads are combined together.

# Chapter 5　Structure Design
# 第 5 章　结构设计

## 5.1　Science and Technology　科学与技术

There is a fundamental difference between science and technology. Engineering or technology is the making of things that did not previously exist, whereas science is the discovering of things that have long existed. Technological results are forms that exist only because people want to make them, whereas scientific results are information of what exists independently of human intentions. Technology deals with the artificial, science with the natural.

科学与技术存在本质的不同。工程技术是研究从未出现过的问题；而科学则是对已长期存在的现象的研究。工程技术的结果是唯一存在的，因为人们预期得到这样的结果；而科学的结果是不以人类的意志为转移的信息。技术解决的是人造的问题；而科学研究的是自然科学。

## 5.2　Structural Engineering　结构工程

Structural engineers are responsible for the detailed analysis and design.
结构工程师负责解决详细的分析与设计问题。

Architectural structures: Buildings, houses, factories. They must work in close cooperation with an architect who will ultimately be responsible for the design.
建筑结构：工业与民用建筑。需与负责最终设计的建筑师紧密合作。

Civil Infrastructures: Bridges, dams, pipelines, offshore structures. They work with transportation, hydraulic, nuclear and other engineers. For those structures they play the leading role.
民用基础设施：桥梁、大坝、管线工程、海洋工程结构。因此与交通、水运、核工业及其他工程师联系密切。在这些结构中他们起主导师作用。

Aerospace, mechanical, naval structures: airplanes, spacecrafts, cars, ships, submarines to ensure the structural safety of those important structures.
航天、力学、航海等：航天器、空间站、汽车、轮船、水下工程等，需要确保该类重要工程的结构安全性。

## 5.3 Structures and their Surroundings  结构及环境因素

Structural design is affected by various environmental constraints:
结构设计受各种环境因素的影响：

(1) Major movements: for example, elevator shafts are usually shear walls which are good at resisting lateral load (wind, earthquake).
主要构件：例如，电梯井筒主要是钢筋混凝土剪力墙结构，其抗侧向力作用好(抗风、抗震)。

(2) Sound and structure interact:
(i) dome roof will concentrate the sound;
(ii) dish roof will diffuse the sound.
声与结构相互作用：穹顶结构利于吸声；碟状屋顶对声波有扩散作用。

(3) natural light:
(i) at roof in a building may not provide adequate light, folded plate will provide adequate lighting;
(ii) bearing and shear wall building may not have enough openings for daylight, frame design will allow more light in.
自然光：
建筑的屋顶可能无法提供足够的光线，折板屋顶则可提供足够的光照；承重墙及剪力墙可能没有充足的开口来采光，而框架结构则不存在这样的问题。

(4) Conduits for cables (electric, telephone, computer), HVAC ducts, may dictate type of floor system.
各种管道(电缆线、电话线、网线等)、HVAC 管道均会影响楼面结构的类型。

(5) Net clearance between columns (unobstructed surface) will dictate type of framing.
柱间距(无障碍的表面)会影响框架结构的布置。

## 5.4 Architecture and Engineering  建筑与结构

Architecture must be the product of a creative collaboration of architects and engineers.
建筑设计应是建筑与工程创造性合作的产品。

Architect stress the overall, rather than elemental approach to design. In the design process, they conceptualize a space-form scheme as a total system. They are generalists.

The engineer, partly due to his/her education think in reverse, starting with details and without sufficient regards for the overall picture. He/She is a pragmatist who knows everything about nothing.

Thus there is a conceptual gap between architects and engineers at all levels of design.
建筑设计注重全局，而非单一元素的设计。在设计过程中，建筑师更注重从整体统一考虑进行概念设计，他们是通才。

而结构工程师则因其专业背景而恰恰相反,其工作一开始即为细部的设计而不必进行总体考虑。因此他(她)们是探求未知领域的现实主义者。

因此,在所有层级的设计中,建筑师和工程师之间都存在概念上的差异。

Engineer's education is more specialized and in depth than the architect's. However, engineer must be kept aware of overall architectural objective.

In the last resort, it is the architect who is the leader of the construction team, and the engineers are his/her servant.

因此,工程师所受的教育比建筑师更专业、更有深度。然而,工程师必须通晓建筑方面的知识。

换言之,建筑师是建设团队的领导人,而结构工程师则担任服从的角色。

## 5.5　Architectural Design Process　建筑设计过程

Schematic: conceptual overall space-form feasibility of basic schematic options. Collaboration is mostly between the owner and the architect.

规划:对基本设计要有全盘的掌握,是业主与建筑师密切合作的成果。

Preliminary: establish basic physical properties of major subsystems and key components to prove design feasibility. Some collaboration with engineers is necessary.

初步设计:完成基本的、主要的分项的设计,证明设计的可行性,与工程师的一些合作是必要的。

Final design: final in-depth design refinements of all subsystems and components and preparation of working documents (blue-prints). Engineers play a leading role.

最终设计:详细、深入的完善所有分项设计成果,准备工作文件(蓝皮书),这个阶段工程师起到关键的作用。

## 5.6　Architectural Design　建筑设计

Architectural design must respect various constraints:

建筑设计需考虑各方面的限制因素:

Functionality: influence of the adopted structure on the purposes for which the structure was erected.

功能性:采用的结构体系对竣工工程使用功能的影响。

Aesthetics: the architect often imposes his/her aesthetic concerns on the engineer. This in turn can place severe limitations on the structural system.

观赏性:建筑师经常将其设计的美学理念强加给土木工程师;这相应地对结构体系造成了严重的限制。

Economy: it should be kept in mind that the two largest components of a structure are labors and materials. Design cost is comparatively negligible.

经济指标:应认识到,工程结构资金投入的两大部分为劳动力及材料,设计费相对来说非常少。

Buildings may have different functions:
工程结构具备不同的功能：

Residential: housing, which includes low-rise (up to 2-3 floors), mid-rise (up to 6-8 floors) and high rise buildings.
民用建筑：住宅类：低层(2～3层)，多层(6～8层)或高层建筑。

Commercial: offices, retail stores, shopping centers, hotels, restaurants.
商业建筑：办公楼、零售商店，商业中心，旅馆，饭店。

Industrial: warehouses, manufacturing.
工业建筑：仓库，工厂等。

Institutional: schools, hospitals, prisons, church, government buildings.
公共结构：学校，医院，监狱，教堂，政府大楼等。

Special: towers, stadium, parking, airport, etc.
特种结构：塔，体育馆，停车场，机场等。

## 5.7　Structural Analysis　结构分析

Given an existing structure subjected to a certain load determine internal forces (axial, shear, flexural, torsional; or stresses), deflections, and verify that no unstable failure can occur.
结构分析是计算给定的结构在承受一定的荷载作用下所产生的内力(弯矩、剪力、弯曲、扭矩或应力)及变形，以确保不发生失稳破坏。

Thus the basic structural requirements are:
因此基本的建筑结构要求是：

Strength: stresses should not exceed critical values.
强度：应力应不超过极限应力值。

Stiffness: deflections should be controlled.
刚度：变形不超过允许限值。

Stability: buckling or cracking should also be prevented.
稳定性：应避免发生屈曲破坏。

## 5.8　Structural Design　结构设计

Given a set of forces, dimension the structural element.
结构设计是给定荷载，确定构件的尺寸。

Steel/wood Structures: Select appropriate section.
钢/木结构需选择合适截面的杆件。

Reinforced Concrete: Determine dimensions of the element and internal reinforcement (number and sizes of reinforcing bars).

钢筋混凝土需确定构件的截面尺寸(钢筋直径及数量)。

For new structures, iterative process between analysis and design is performed. A preliminary design is made using rules of thumbs (best known to engineers with design experience) and analyzed. Following design, we check the following:

对于一新型建筑，其结构分析与结构设计是交互进行的。初步设计是设计工程师根据经验来完成的。根据设计需要检查以下几项：

(1) Serviceability: deflections, crack widths under the applied load. Compare with acceptable values specified in the design code.

实用性：在荷载作用下所产生的变形、裂缝的宽度应满足规范的相应要求。

(2) Failure (limit state): and compare the failure load with the applied load times the appropriate factors of safety.

极限状态：根据破坏荷载及施加的荷载，确定合适的安全系数。

If the design is found not to be acceptable, then it must be modified and reanalyzed.

若某一方面不满足要求，需进行重新分析和设计。

For existing structures rehabilitation, or verification of an old infrastructure, analysis is the most important component. In summary, analysis is always required.

对已建结构的加固，或验证某一基础设施是否满足要求，则结构分析是最重要的组成部分。总而言之，结构分析始终是不可或缺的。

## 5.9  Load Transfer Mechanisms  荷载传递机理

From strength of materials, loads can be transferred through various mechanisms, as shown in Figure 5.1.

根据材料强度，荷载可根据不同途径进行传递。如图 5.1 所示。

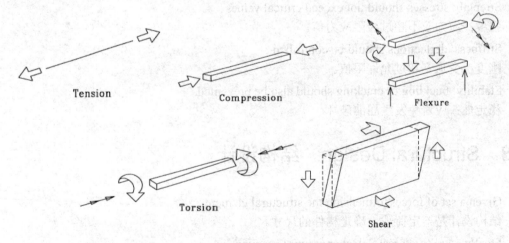

Figure 5.1  Types of Forces in Structural Elements

图 5.1  构件中力的分类

Tension—张力(拉力)；Compression—压力；Flexure—弯曲力(挠力)；Torsion—扭力；Shear—剪力

Axial: cables, truss elements, arches, membrane, shells.
轴向受力构件：索、桁架、拱、膜及壳结构。
Flexural: Beams, frames, grids, plates.
弯曲受力构件：梁、框架、柱、板。
Torsional: Grids, 3D frames.
扭转受力构件：梁、空间框架。
Shear: Frames, grids, shear walls.
剪切受力构件：框架、梁及剪力墙。

## 5.10　Structure Types　结构承重体系

Structures can be classified as follows:
结构体系可作如下分类：

Tension & Compression Structures: no shear, flexure or torsion. Those are the most efficient types of structures.
拉、压结构：无剪切、弯曲及挠曲效应，是材料利用率最好的构件。

Cable (tension only): The high strength of steel cables, combined with the efficiency of simple tension, makes cables ideal structural elements to span large distances such as bridges and dish roofs, as shown in Figure 5.2. A cable structure develops its load carrying capacity by adjusting its shape so as to provide maximum resistance (form follows function). Care should be exercised in minimizing large deflections and vibrations.

索结构(张拉结构)：钢索强度大；通过有效的张拉力的施加可使索结构能够跨越很大的距离，如在桥梁结构及大跨屋盖结构中的应用，如图 5.2 所示。索结构可通过改变其形状来获得最大的承载能力(形式取决于功能)。需要注意的是保证出现最小的下垂变形及振动效应。

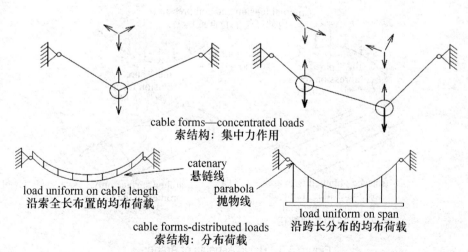

Figure 5.2　Basic Aspects of Cable Systems
图5.2　索结构

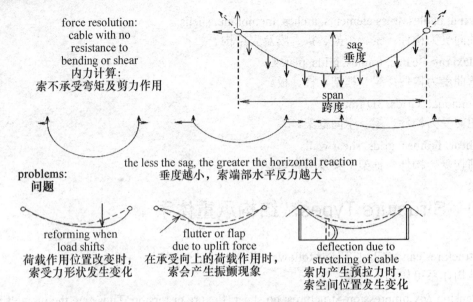

Figure 5.2　Basic Aspects of Cable Systems(Continued)

图 5.2　索结构(续)

Arches (mostly compression) is a reversed cable structure. In an arch, flexure/shear is minimized and most of the load is transferred through axial forces only. Arches are used for large span roofs and bridges, as shown in Figure 5.3.

拱结构(大部分只承受压力)是与索结构相反的结构类型。在拱结构中，弯曲及剪切效应最小，大部分荷载只通过拱内轴力来传递。拱结构广泛应用于大跨屋盖结构或长跨桥梁中，如图 5.3 所示。

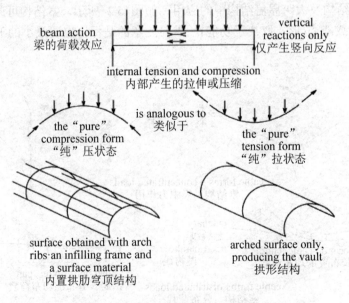

Figure 5.3　Basic Aspects of Arches

图 5.3　拱的基本结构

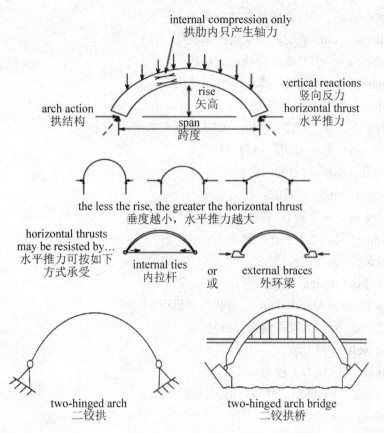

Figure 5.3　Basic Aspects of Arches(Countinued)

图5.3　拱的基本结构(续)

# New Words and Expressions　生词和短语

1. Structural Engineering　结构工程；(桥梁等的)结构工程学；建筑工程
2. architectural structure　建筑结构；总体结构
3. Civil Infrastructure　民用基础设施
4. pipeline　n. 管道；输油管；传递途径
5. offshore structure　近海结构；海域结构物
6. hydraulic　adj. 液压的；水力的；水力学的
7. constraint　n. 约束；限制；约束条件
8. conduit　n. [电] 导管；沟渠；导水管
9. generalist　n. (有多方面知识和经验的)通才；多面手
10. pragmatist　n. 实用主义者；爱管闲事的人
11. hierarchical　adj. 分层的；等级体系的
12. schematic　n. 原理图；图解视图

13. collaboration    *n.* 合作；勾结；通敌
14. aesthetic    *adj.* 美的；美学的；审美的
15. warehouse    *n.* 仓库；货栈；大商店
16. stadium    *n.* 体育场；露天大型运动场
17. axial    *adj.* 轴的；轴向的
18. shear    *n.* [力] 切变；修剪；大剪刀
         *vi.* 剪；剪切；修剪
19. flexural    *adj.* 弯曲的；曲折的
20. torsional    *adj.* 扭转的；扭力的
21. deflection    *n.* 偏向；挠曲；偏差
22. buckling    *n.* [力] 屈曲；膨胀
23. cracking    *n.* 破裂；裂化
24. preliminary design    初步设计
25. Load Transfer Mechanism    荷载传递机理
26. shell    *n.* 壳，贝壳；炮弹；外形
27. shear wall    剪力墙
28. tension    *n.* 张力，拉力
29. vibration    *n.* 振动；犹豫；心灵感应
30. parabola    *n.* 抛物线

# Exercises    练习

Ⅰ. Write a T in front of a statement if it is true according to the text and write an F if it is false.

1. There is a fundamental difference between science and and technology.

2. The major movements of elevator shafts are shear walls which are good at resisting gravity load.

3. Architect stress the overall, rather than elemental approach to design.

4. Architectural design is hierarchical and it can be divided into four sections.

5. A cable structure develops its load carrying capacity by adjusting its shape so as to provide maximum resistance.

Ⅱ. Complete the following sentences.

1. _____ must work in close cooperation with an architect who will ultimately be responsible for the design.

2. Architecture must be the product of a creative collaboration of _____.

3. It should be kept in mind that the two largest components of a structure are

# Structure Design

_____.

4. The internal forces of an existing structure which should be analyzed are _____, _____, _____, _____, _____.

5. In an arch, _____ is minimized and most of the load is transfered through _____ only.

Ⅲ. Translate the following sentences into Chinese.

1. Engineering or technology is the making of things that did not previously exist, whereas science is the discovering of things that have long existed.

2. Elevator shafts are usually shear walls which are good at resisting lateral load.

3. Architecture must be the product of a creative collaboration of architects and engineers.

4. Given an existing structure subjected to a certain load determine internal forces, deflections, and verify that no unstable failure can occur.

5. A cable structure develops its load carrying capacity by adjusting its shape so as to provide maximum resistance.

Ⅳ. Translate the following sentences into English.

1. 科学和技术之间有一个根本的区别。
2. 结构工程师对细部的分析和建筑结构的设计负责。
3. 建筑师应该强调整体，而不是强调主要的近似设计。
4. 如果设计不被接受，那么它必须被修改和再分析。
5. 在一个拱中，弯/剪是最小的，而大部分荷载只通过轴向力转移。

# Answers　答案

Ⅰ.

1. T　2. F　3. T　4. F　5. T

Ⅱ.

1. Architectural structure
2. architects and engineers
3. labors and materials
4. axial, shear, flexural, torsional, stresses
5. flexure/shear , axial forces

Ⅲ.

1. 工程或技术是制作以前不存在的事物，而科学是发现早就存在了的东西。
2. 电梯井通常能够很好地抵抗水平荷载的受力剪力墙。
3. 建筑必须是建筑师和工程师的创造性合作的产品。

4. 给现有的结构施加一定荷载以确定内力和变形，并确保没有不稳定的故障发生。

5. 电缆结构通过调整其形状发展其承载能力，以提供最大的阻力。

IV.

1. There is a fundamental difference between science and technology.

2. Structural engineers are responsible for the detailed analysis and design of architectural structures.

3. Architect should stress the overall, rather than elemental approach to design.

4. If the design is found not to be acceptable, then it must be modified and reanalyzed.

5. In an arch, flexure/shear is minimized and most of the load is transferred through axial forces only.

# Chapter 6  A Brief History of Structural Architecture
# 第 6 章  结构建筑发展简史

If I have been able to see a little farther than some others, it was because I stood on the shoulders of giants.

—Sir Isaac Newton

如果我看得比别人更远一点，那是因为我站在了巨人的肩膀上。

——牛顿

More than any other engineering discipline, Architecture/Mechanics/Structures is the proud outcome of a long and distinguished history. Our profession, second oldest, would be better appreciated if we were to develop a sense of our evolution.

与任何其他的工程学科相比，建筑/机械/结构专业拥有悠久且辉煌的历史，其发展成果是值得骄傲的。我们所从事的专业是第二古老的专业，若能发展出创新的观念，会更值得人们欣赏。

## 6.1  Before the Greeks  古希腊时期

Throughout antiquity, structural engineering exists as an art rather than a science. No record exists of any rational consideration, either as to the strength of structural members or as to the behavior of structural materials. The builders were guided by rules of thumbs and experience, which were passed from generation to generation, guarded by secrets of the guild, and seldom supplemented by new knowledge. Despite this, structures erected before Galileo (pyramids, Via Appia, aqueducts, Coliseums, Gothic cathedrals) are by modern standards quite phenomenal.

在古代，结构工程作为一门艺术而非科学存在。无论是对结构构件的强度或对材料的力学行为均没有记录表明存在任何理性的思考。建造者在代代相传的经验法则的指导下工作，行业协会保守着这些规则和经验，很少补充新的知识。尽管如此，在伽利略之前的结构(像金字塔，亚壁古道，管路，竞技场，哥特式大教堂)按现代标准评价其成就也是相当惊人的。

The first structural engineer in history seems to have been Imhotep, one of only two commoners to be died. He was the builder of the step pyramid of Zoser, Sakkara about 3000 B.C., and yielded great influence over ancient Egypt.

历史上第一个结构工程师似乎是伊姆霍特普,他是仅有的死后有公地使用权的两个人之一。在公元前 3000 年左右,他负责建起了萨卡拉的昭塞尔金字塔,并在古埃及产生了巨大的影响。

*Hammurabi's code in Babylonia* (1750 BC) (as shown in Figure 6.1) included among its 282 laws penalties for those architects whose houses collapsed.

巴比伦王国的汉谟拉比法典(公元前 1750 年)(如图 6.1 所示),包括 282 项法律,用于惩罚那些房屋倒塌的建筑师。

Figure 6.1　Hammurabi's Code

图 6.1　汉谟拉比法典

## 6.2　Greeks　希腊时期

The Greek philosopher Pythagoras (born around 582 B.C.) founded his famous school, which was primarily a secret religious society, at Crotona in southern Italy. At his school he allowed neither textbooks nor recording of notes in lectures, on pain of death. He taught until the age of 95, and is reported to have coined the term mathematics which means literally the "science of learning" (and also the word philosopher meaning "one who loves wisdom").

古希腊哲学家毕达哥拉斯(生于公元前 582 年左右)在意大利南部的克罗多尼创立了著名的学校,主要是一个秘密的宗教组织。在他的学校,不允许讲教科书或在讲座中记录有关死亡的痛苦。他一直教到 95 岁,据记载,他已经创造了"数学"这个术语,其字面意思是"学习科学"(而"哲学家"的意思为"爱智慧")。

Aristotle (384-322 B.C.) was Dean of the Lyceum, a college just outside the city gates of Athens, and was a man of universal ability. He is credited with having written in more than 25 different fields of knowledge, one of the most influential men of early civilization.

亚里士多德(公元前 384—322)是"学园"的院长,这个大学就在雅典城门外,亚里士多德是一个无所不通的人。他因写了超过 25 个不同领域的著作而成为早期文明最具影响力的人物之一。

A pupil of Aristotle was Alexander the Great (356-323 B.C.) who founded the city of Alexandria in 323 B.C. Upon his death, one of his generals Ptolemy I became Pharaoh and established a library.

亚里士多德的学生亚力山大大帝(公元前 356—323)在公元前 323 年创立了城市亚历山

大城。在他死后，他的一个将军托勒密一世成了法老并且建立了一个图书馆。

The library of Alexandria was founded with the private library of Aristotle as a nucleus, and later became the largest of the ancient world, containing about 700,000 scrolls. Many of these scrolls were subsequently brought to the attention of the western world through translations by the Arabs.

亚历山大图书馆的成立以亚里士多德的私人图书馆为核心，并且后来成为古代世界最大的图书馆，含有图书约 70 万卷。其中许多随后被翻译成阿拉伯语得到西方世界的关注。

Alexandria was also the seat of the first university (with a reported enrollment of 14,000 students), and its first professor of geometry was Euclid (315-250 B.C.).

亚历山大也是第一座大学(据记载招了 14 000 名大学生)，它的第一个几何学教授是欧几里得(公元前 315—250 年)。

The greatest of the Greeks was Archimedes (287-212) who was one of the greatest physicists of the ancient world and one of its greatest mathematicians, as shown in Figure 6.2. He is considered by many as the founder of mechanics because of his treatise "On Equilibrium". He introduced the concept of center of gravity.

最伟大的希腊人阿基米德(287—212)是古代世界最伟大的物理学家和数学家之一，如图 6.2 所示。他凭借力学平衡的专著而被许多人认为是力学的创始人。他引入了重心的概念。

Figure 6.2　Archimedes
图 6.2　阿基米德

He refused to write about "practical study" such as machines, catapults, spiral pumps, and others. It was one such invention (the lens) which kept the Roman armies at bay outside Syracuse for three years. When the city fell, he was supposed to have had his life spared. But the circumstances of his subsequent death are obscure. By some accounts he was killed by an ignorant soldier who disobeyed orders, and by other he was slain because he was too busy solving a mathematical problem to appear in front of the Roman consul and conqueror of Syracuse.

他拒绝写有关实践性的内容，如机器、发射器、螺旋泵和其他一些东西。他发明的镜头拒罗马军队于锡拉丘兹湾外三年。当这个城市沦陷时，他本来能够使自己获得赦免。但他其后死亡的原因仍扑朔迷离。据说他是被一个愚蠢的不服从命令的士兵杀害的，另一种说法是他因为忙于解决一个数学问题而没有去见罗马领事和锡拉丘兹的征服者才被杀的。

## 6.3 Romans 罗马时期

Science made much less progress under the Romans than under the Greeks. The Romans apparently were more practical, and were not as interested in abstract thinking though they were excellent fighters and builders.

与希腊人相比，罗马人的科学进步不大，罗马人显然是更实际的，他们的兴趣不在抽象思维，然而他们是优秀的战士和建造者。

As the Roman Empire expanded, the Romans built great roads (some of them still in use) such as the Via Appia; also they built great bridges (such as the third of a mile bridge over the Rhine built by Caesars), and stadium (Coliseum).

随着罗马帝国的扩张，罗马人建造了巨大的道路(其中许多仍在使用)，如亚壁古道；他们也建大桥(如由恺撒建造的横跨在莱茵河上的第三座一英里长的桥)和体育场(罗马斗兽场)。

One of the most notable Roman constructions was the Pantheon, as shown in Figure 6.3. It is the best-preserved major evidence of ancient Rome and one of the most significant buildings in architectural history. In shape it is an immense cylinder concealing eight piers, topped with a dome and fronted by a rectangular colonnaded porch. The great vaulted dome is 43 m (142 ft) in diameter, and the entire structure is lighted through one aperture, called an oculus, in the center of the dome. The Pantheon was erected by the Roman emperor Hadrian between AD 118 and 128.

其中最著名的一个建筑是罗马万神殿，如图 6.3 所示。这是古代罗马保存最好的证据，并且是其历史上最著名的建筑之一。在外形上，它是一个巨大的圆柱状物八墩，顶部有一个圆顶，还有成排矩形带有柱廊的玄关。巨大的拱形的穹顶直径43米(142英尺)，整个建筑通过穹顶中心一个称为眼睛的孔结构进行照明。万神殿建于罗马皇帝哈德良的公元 118 年和 128 年之间。

Figure 6.3　Pantheon
图 6.3　罗马万神殿

Marcus Vitruvius Pollio (70–25 B.C.) was a Roman architect and engineer. He was an artillery engineer in the service of the first Roman emperor, Augustus. His Ten Books on

Architecture (Vitruvius 1960) is the oldest surviving work on the subject and consists of dissertations on a wide variety of subjects relating to architecture, engineering, sanitation, practical hydraulics, acoustic vases, and the like. Much of the material appears to have been taken from earlier extinct treatises by Greek architects. Vitruvius's writings have been studied ever since the Renaissance as a thesaurus of the art of classical Roman architecture, as shown in Figure 6.4.

马库斯·维特鲁威·波利奥(公元前70—25)是古罗马的建筑师和工程师。他是第一个罗马皇帝奥古斯都的炮兵工程师。他的建筑十书(维特鲁威,1960)是现存最古老的著作,包括多个领域的有关建筑工程、卫生、水力学和声学等多种论文。大部分内容似乎是取自早期希腊建筑师写的已经绝迹的论文。维特鲁威的著作研究了自文艺复兴时期作为艺术宝典的古典罗马建筑。如图6.4所示。

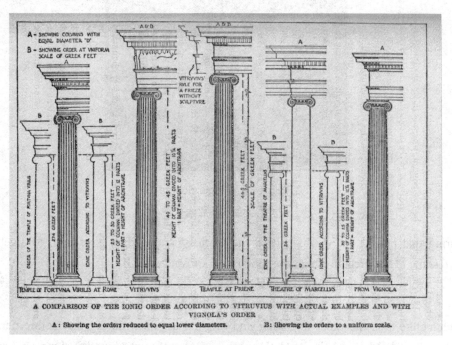

Figure 6.4　From Vitruvius Ten Books on Architecture (Vitruvius, 1960)

图6.4　波利奥的建筑《十书》(维特鲁威,1960)

## 6.4　The Medieval Period (477-1492)
　　　中世纪时期(477—1492)

This period, also called the Dark Ages, was marked by a general decline of civilization. Throughout Europe following the decline and fall of the western Roman Empire.

The eastern Roman Empire on the other hand was to continue, and the center of Greek life had by then been transferred to Constantinople. This city exerted great influence throughout Asia Minor.

这一时期的显著特点是人类文明的普遍下降，也被称为黑暗时代。西罗马帝国的衰亡影响了整个欧洲。另一方面，东罗马帝国继续，并且希腊文明的重心被转移到君士坦丁堡，这个城市在亚洲产生了重大影响。

Hagia Sophia, also Church of the Holy Wisdom, as shown in Figure 6.5, was the most famous Byzantine Structure in Constantinople (now Istanbul). Built by Emperor Justinian Ⅰ, its huge size and daring technical innovations make it one of the world's key monuments. The size of its dome though, 112 ft, was nevertheless smaller than the one of the Pantheon in Rome.

如图 6.5 所示的圣索菲亚大教堂意味着神的智慧，是君士坦丁堡(现在的伊斯坦布尔)最著名的拜占庭结构。它由皇帝查士丁尼一世建造，其庞大的尺寸和大胆的技术创新使其成为世界主要的古迹之一。尽管它的圆顶有 112 英尺，但仍然小于罗马的万神殿。

Figure 6.5　Hagia Sophia
图 6.5　圣索菲亚大教堂

During that period, the Arabs carried the torch of knowledge, gave birth to algebra, and translated some of the great books of the Library of Alexandria.

在这一期间，阿拉伯人传递了知识的火炬，产生了代数，翻译了亚历山大图书馆的一些巨著。

Architecture was the most important and original art form during the Gothic period, the principal structural characteristics of Gothic architecture arose out of medieval masons' efforts to solve the problems associated with supporting heavy masonry ceiling vaults over wide spans. The problem was that the heavy stonework of the traditional arched barrel vault and the groin vault exerted a tremendous downward and outward pressure that tended to push the walls upon which the vault rested outward, thus collapsing them. A building's vertical supporting walls thus had to be made extremely thick and heavy in order to contain the barrel vault's outward thrust.

建筑是哥特时期最重要和最原始的艺术形式。哥特式建筑的主要结构特点要求中世纪的石匠努力解决大跨度重型砌体天花板穹顶的相关问题。传统的筒形拱和拱顶石的自重产生了巨大的向下和向外的压力，往往把墙壁上的拱顶向外推并造成破坏。因此建筑物的垂直支撑墙必须非常厚重以抵消筒形拱顶向外的推力。

Medieval masons solved this difficult problem about 1120 AD with a number of brilliant innovations. First and foremost, they developed a ribbed vault, in which arching and intersecting stone ribs support a vaulted ceiling surface that is composed of mere thin stone panels. This greatly reduced the weight (and thus the outward thrust) of the ceiling vault, and since the vault's weight was now carried at discrete points (the ribs) rather than along a continuous wall edge, separate widely spaced vertical piers to support the ribs could replace the continuous thick walls. The round arches' barrel vault were replaced by pointed (Gothic) arches which distributed thrust in more directions downward from the topmost point of the arch.

中世纪的石匠在1120年利用许多富有智慧的创新解决了这一难题。第一点且是最重要的一点，他们开发了一种带肋拱顶，这种拱顶由相交的石肋支撑在由薄石板组成的拱形天花板的表面。这大大降低了天花板拱顶的重量(从而减小了向外的推力)。由于拱顶的重量开始沿离散点(肋骨)而不是沿着一个连续的墙的边缘传递，单独的广泛隔开的垂直的柱子代替了连续的厚墙壁。圆拱的桶形拱顶被尖拱(哥特式)代替，这种结构使拱推力从拱的最高点沿多个方位向下分布。

Since the combination of ribs and piers relieved the intervening vertical wall spaces of their supportive function, these walls could be built thinner and could even be opened up with large windows or other glazing. A crucial point was that the outward thrust of the ribbed ceiling vaults was carried across the outside walls of the nave, first to an attached outer buttress and then to a freestanding pier by means of a half arch known as a buttress. The buttress leaned against the upper exterior of the nave (thus counteracting the vault's outward thrust), crossed over the low side aisles of the nave, and terminated in the freestanding buttress pier, which ultimately absorbed the ceiling vault's thrust. These elements enabled Gothic masons to build much larger and taller buildings than their Romanesque predecessors and to give their structures more complicated ground plans. The skillful use of buttresses made it possible to build extremely tall, thin-walled buildings whose interior structural system of columnar piers and ribs reinforced an impression of soaring verticality.

由于肋骨和墩的组合结构减轻了垂直壁的支撑功能，这些墙壁可以建造得更薄，甚至可以开大窗户或其他玻璃窗。关键的一点是，天花板的肋拱顶向外的推力作用在殿的外墙上，随后加载在外扶壁上，然后通过一个独立的半拱靠在外殿上(因此抵消了这种拱顶的外在推力)，再穿过低侧通道的正厅，最终吸收了天花板拱顶的推力。这些力学元素使得哥特式的石匠能够比他们的前辈建造更大、更高和更复杂的罗马式建筑。支墩的巧妙运用使得建造墙壁极高极薄的建筑成为可能。内部的柱墩和肋使得建筑给人以飞腾向上的印象。

## 6.5　The Renaissance　文艺复兴时期

During the Renaissance there was a major revival of interest in science and art.
在文艺复兴时期主要是科学和艺术的复兴。

## 6.5.1　Leonardo da Vinci　列奥纳多·达·芬奇

Leonardo da Vinci(1452-1519) was the most outstanding personality of that period (and of human civilization for that matter). He was not only a great artist (*Mona Lisa*), but also a great scientist and engineer.

列奥纳多·达·芬奇(1452—1519)是那个时代(人类文明)最优秀的人物。他不仅是一位伟大的艺术家(《蒙娜丽莎》)，也是一位伟大的科学家和工程师。

He did not write books, but much information was found in his notebooks, one of the most famous (Codex Leicester) was recently purchased by Bill Gates whose company Corbis made a CD-ROM from it. He was greatly interested in mechanics, (Timoshenko 1982), and in one of his notes he states "Mechanics is the Paradise of mathematical science because here we come to the fruits of mathematics." He was the first to explore concepts of mechanics, since Archimedes, using a scientific approach. He applied the principle of virtual displacements to analyze various systems of pulleys and levers. He appears to have developed a correct idea of the thrust produced by an arch.

他从不著书，但他的笔记包含大量的信息，最著名的莱斯特抄本最近由比尔·盖茨的公司 Corbis 购买并做成了光盘。达·芬奇对力学很感兴趣(铁木辛柯，1982)，在笔记中他写道："力学是数学科学的天堂，因为我们在这里得到数学的成果。" 他是自阿基米德以来第一个用科学的方法探索力学概念的人。他运用虚位移原理分析滑轮和杠杆的不同系统。他似乎对拱产生推力已经形成了正确的观点。

Leonardo also studied the strength of structural materials experimentally. He tried to determine the tensile strength of an iron wire of different length.

列奥纳多还研究了结构材料的强度试验研究。他试图确定不同长度的铁丝的抗拉强度。

He also studied the load carrying capacity of a simply supported uniformly loaded beam and concluded that "the strength of the beam supported at both ends varies inversely as the length and directly as the width".

他还研究了均匀加载简支梁的承载能力，得出的结论是，"两端支承简支梁的强度与梁的长度成反比，与宽成正比"。

## 6.5.2　Brunelleschi　布鲁内莱斯基

Brunelleschi(1377-1446) was a Florentine architect and one of the initiators of the Italian Renaissance. His revival of classical forms and his championing of an architecture based on mathematics, proportion, and perspective make him a key artistic figure in the transition from the Middle Ages to the modern area.

布鲁内莱斯基(1377—1446)是佛罗伦萨的建筑师，并且是意大利文艺复兴的倡导者之一。他对于古典形式的复兴以及他基于数学、比例和角度等的建筑，使他成为一个从中世纪向现代过渡时期的关键人物。

He was born in Florence in 1377 and received his early training as an artisan in silver and gold. In 1401 he entered, and lost, the famous design competition for the bronze doors of the Florence Baptistery. He then turned to architecture and in 1418 received the commission to execute the dome of the unfinished Gothic Cathedral of Florence. The dome, as shown in Figure 6.6, is a great innovation both artistically and technically, consists of two octagonal vaults, one inside the other. Its shape was dictated by its structural needs. It's one of the first examples of architectural functionalism.

他 1377 年出生在佛罗伦萨，他早期接受的训练是成为一个金银工匠。1401 年他参加了在佛罗伦萨著名的青铜门设计竞赛，却名落孙山。之后他转向建筑设计，在 1418 年受委托设计那些未完成的哥特式佛罗伦萨大教堂的圆顶。如图 6.6 所示的圣母百花大教堂的穹顶是伟大的艺术和技术创新，由两个八角形的拱顶一个套一个地组成。结构的需要决定了它的形状。它是最早体现建筑的实用功能主义的建筑之一。

Figure 6.6 Florence's Cathedral Dome
图 6.6 圣母百花大教堂

Brunelleschi made a design feature of the necessary eight ribs of the vault, carrying them over to the exterior of the dome, where they provide the framework for the dome's decorative elements, which also include architectural reliefs, circular windows. This was the first time that a dome created the same strong effect on the exterior as it did on the interior.

布鲁内莱斯基对拱顶采用八肋的设计特点，并将它们延伸到拱顶的外部，为屋顶提供了不同的装饰元素，这也包括建筑浮雕，圆形的窗户。这是首次拱顶在内外部都给人强烈的视觉冲击。

## 6.5.3　Galileo　伽利略

Galileo (1564-1642, as shown in Figure 6.7) was born in Pisa in 1564. He received his early education in Latin, Greek and logic near Florence. In 1581 he entered the University of Pisa to study medicine, but he soon turned to philosophy and mathematics, leaving the university without a degree in 1585. For a time he tutored privately and wrote on hydrostatics and natural motions, but he did not publish.

伽利略(1564—1642，如图 6.7 所示)在 1564 年生于比萨。他在佛罗伦萨附近接受了有关拉丁语、希腊语和逻辑的早期教育。1581 年伽利略进入比萨大学学习医学，但他很快转向哲学与数学，在没有拿到学位的情况下于 1585 年退学。一段时间内他自学了流体静力学和自然的运动，但他并没有公开发表。

Figure 6.7　Galileo
图 6.7　伽利略

In 1589 he became professor of mathematics at Pisa, where he is reported to have shown his students the error of Aristotle's belief that speed of fall is proportional to weight, by dropping two objects of different weight simultaneously from the Leaning Tower, thus modern dynamics was born. His contract was not renewed in 1592, probably because he contradicted Aristotelian professors. The same year, he was appointed to the chair of mathematics at the University of Padua, where he remained until 1610.

1589 年伽利略在比萨成为数学教授，据说他在那里向他的学生演示了亚里士多德定律的错误：物体下降速度与重量成正比。通过两个不同重量的物体同时从斜塔下落的试验，产生了现代动力学。他的合同在 1592 年没有被续签，可能是因为他反驳亚里士多德派的教授。同年，他被任命为帕多瓦大学数学系的教授并在那里工作到 1610 年。

In Padua he achieved great fame, and lecture halls capable of containing 2,000 students from all over Europe were used. He subsequently became interested in astronomy and built one of the first telescope through which he saw Jupiter and became an ardent proponent of the Copernican theory (which stated that the planets circle the sun as opposed to the Aristotelian and Ptolemaic assumptions that it was the sun which was circling Earth). This theory being condemned by the church, he received a serious warning to avoid theology and limit himself to physical reasoning.

在帕多瓦伽利略取得了巨大的声誉，在容纳欧洲各地 2000 名学生的演讲厅演讲。他后来又对天文学感兴趣并建成了第一个望远镜，通过望远镜他观察到了木星并成为哥白尼理论的支持者(这说明行星围绕太阳运转，但亚里士多德和托勒密的假设是太阳绕地球运转)。这一理论受到教会的谴责，他被严厉警告避开神学并被限制进行物理推理研究。

Galileo's lifelong struggle to free scientific inquiry from restriction by philosophical and theological interference stands beyond science. Since the full publication of Galileo's trial documents in the 1870s, entire responsibility for Galileo's condemnation has customarily been placed on the Roman Catholic Church. In October 1992 a papal commission acknowledged the Vatican's error.

伽利略一生通过哲学为科学自由而奋斗，使科学免于神学的干扰。自从对伽利略的审判文件在 1870 年完全公开以来，给伽利略错误定罪的整个责任都落在罗马天主教会身上。1992 年 10 月罗马教皇委员会承认了梵蒂冈的错误。

## 6.6 Pre Modern Period, Seventeenth Century
新世纪早期，十七世纪时期

### 6.6.1 Hooke 胡克

Hooke(1635-1703) was best known for his study of elasticity but also original contributions to many other fields of science.

胡克(1635—1703)最为人所知的是他在弹性力学方面的研究，同时，他在其他的许多科学领域也做出了原创性的贡献。

Hooke was born on the Isle of Wight and educated at the University of Oxford. He served as assistant to the English physicist Robert Boyle and assisted him in the construction of the air pump. In 1662 Hooke was appointed curator of experiments of the Royal Society and served in this position until his death. He was elected a fellow of the Royal Society in 1663 and was appointed Gresham Professor of Geometry at Oxford in 1665. After the Great Fire of London in 1666, he was appointed surveyor of London, and he designed many buildings.

胡克出生于怀特岛上，在牛津大学受过教育。他担任过英国物理学家罗伯特·波义耳的助手，协助他的空气泵施工。1662 年胡克被任命为英国皇家学会的实验室主任，直到他去世，他一直担任此职。他在 1663 年被选为皇家学会会员，1665 年被任命为牛津大学的格雷沙姆几何学教授。1666 年伦敦大火之后，他成为伦敦城的测绘工程师并设计了许多建筑。

Hooke anticipated some of the most important discoveries and inventions of his time but failed to carry many of them through to completion. He formulated the theory of planetary motion as a problem in mechanics, and grasped, but did not develop mathematically.

胡克预见了他那个时期一些最重要的发现和发明，但却并未将它们完成。他发现了行星运动原理，但并没有将这个力学问题归结为数学问题。

His most important contribution was published in 1678 in the paper *De Potentia Restitutiva*. It contained results of his experiments with elastic bodies, and was the first paper in which the elastic property of material was discussed, as shown in Figure 6.8.

他最重要的贡献是于 1678 年发表在《论弹簧》一文中关于物体弹性方面的实验结果。这是世界上第一篇讨论材料弹性性能的论文。如图 6.8 所示。

Figure 6.8　Hooke and the Experimental Set up Used by Hooke

图 6.8　胡克及其实验装置

Take a wire string of 20, or 30, or 40 ft long, and fasten the upper part thereof to a nail, and to the other end fasten a scale to receive the weights. Then with a pair of compasses take the distance of the bottom of the scale from the ground or underneath, and set down the said distance, then put in weights into the said scale and measure the several stretching of the said string, and set them down. Then compare the several stretching of the said string, and you will see that they will always bear the same proportion one to the other that the weights do that made them.

将长各为 20、30、40 英尺的金属线一端用钉子固定，而另一端系一天平来进行加载，然后用圆规测量天平底部到地面的距离，再把荷载放到天平上分别测量金属线的伸长，放下荷载。然后比较几种拉伸作用下的不同的金属线，会得出金属线的伸长与它们加载的重量总是成同一比例的结论。

Because he was concerned about patent rights to his invention, he did not publish his law when first discovered it in 1660.

因担心发明专利权的问题，胡克没有在 1660 年第一时间公布他的发现。

## 6.6.2　Newton　牛顿

Born on Christmas day in the year of Galileo's death, Newton, Figure 6.9 was Professor of Mathematics at Cambridge University.

牛顿(1642—1727)出生于伽利略去世那一年的圣诞节。如图 6.9 所示是在剑桥大学当数学教授的牛顿。

Figure 6.9　Isaac Newton

图 6.9　牛顿

In 1684 Newton's solitude was interrupted by a visit from Edmund Halley, the British astronomer and mathematician, who discussed with Newton the problem of orbital motion. Newton had also pursued the science of mechanics as an undergraduate, and at that time he had already entertained basic notions about universal gravitation. As a result of Halley's visit, Newton returned to these studies.

1684 年，牛顿的与世隔绝因为埃德蒙·哈雷的到访而被打破，埃德蒙·哈雷是英国天文学家和数学家，他与牛顿讨论了轨道运动的问题。牛顿作为本科生曾经学习力学这门科学，在那个时候他已经有了关于万有引力的基本概念。由于哈雷的拜访，牛顿重新开始了这些研究。

During the following two and a half years, Newton established the modern science of dynamics by formulating his three laws of motion. Newton applied these laws to Kepler's laws of orbital motion formulated by the German astronomer Johannes Kepler and derived the law of universal gravitation. Newton is probably best known for discovering universal gravitation, which explains that all bodies in space and on earth are accepted by the force called gravity. He published this theory in his book *Philosophiae Naturalis Principia Mathematica* or simply *Principia*, in 1687, as shown in Figure 6.10. This book marked a turning point in the history of science.

在接下来的两年半的时间里，通过提出三大运动定律，牛顿建立了现代的牛顿动力学。牛顿应用这些定律研究德国天文学家开普勒提出的行星轨道运动并提出了万有引力。牛顿一生最出名的是发现了万有引力，这解释了在宇宙中或地球上的所有物体都要承受类似于重力的作用。1687 年他发表了《自然哲学的数学原理》一书，或简称《原则》，如图 6.10 所示。这本书为科学史上的转折点。

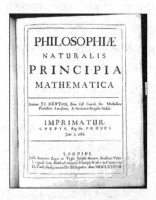

Figure 6.10  Philosophiae Naturalis Principia Mathematica, Cover Page
图 6.10  自然哲学的数学原理，封面

Newton also engaged in a violent dispute with Leibniz over priority in the invention of calculus. Newton used his position as president of the Royal Society to have a committee of that body investigate the question, and he secretly wrote the committee's report, which charged Leibniz with deliberate plagiarism. Newton also compiled the book of evidence that the society published. The elects of the quarrel lingered nearly until his death in 1727.

牛顿还与莱布尼茨就谁是微积分的开创者进行过激烈的争论。莱布尼茨在微积分的研究方面处于领先地位。牛顿作为英国皇家学会主席拥有一个委员会解决这个问题。他偷偷写了委员会的报告指控莱布尼茨故意剽窃。牛顿还编辑并利用学会出版了书作为证据。他们之间的争吵一直持续到将近1727年牛顿去世为止。

In addition to science, Newton also showed an interest in alchemy, mysticism, and theology. Many pages of his notes and writings particularly from the later years of his career are devoted to these topics. However, historians have found little connection between these interests and Newton's scientific work.

除了自然科学，牛顿也对炼金术、神秘主义和神学感兴趣。他晚年的许多笔记和著作主要是关于这些主题的。但是，历史学家发现，这些兴趣与牛顿在自然科学方面的研究联系不大。

## 6.7  The Pre-Modern Period: Coulomb and Navier 新世纪早期：库仑，纳维

Coulomb (1736-1806) was a French military engineer, as shown in Figure 6.11. He was the first to publish the correct analysis of the stresses in fixed beam with rectangular cross section. He used Hooke's law, placed the neutral axis in its exact position, developed the equilibrium of forces on the cross section with external forces, and then correctly determined the stresses. Coulomb did also research on magnetism, friction, and electricity. In 1777 he invented the torsion balance for measuring the force of magnetic and electrical attraction. With this invention, Coulomb was able to formulate the principle, now known as Coulomb's law, governing the interaction between electric charges. In 1779 Coulomb published the treatise theories *des machines simples* (Theory of Simple Machines), an analysis of friction in machinery. After the war Coulomb came out of retirement and assisted the new government in devising a metric system of weights and measures. The unit of quantity used to measure electrical charges, the Coulomb, was named for him.

库仑(1736—1806)是一个法国军事工程师，如图6.11所示。他是第一个发表了正确分析矩形截面梁横截面应力分布规律的人。他用胡克定律，找到了中性轴的精确位置，形成了横截面上外力与内力的平衡理论，然后正确地计算了应力。库仑也在摩擦产生电和磁方面做了研究。1777年他发明了用于测量磁性和电引力的扭秤。凭借这个发明，库仑的发现得以形成理论，也就是所谓的计算电荷之间相互作用的库仑定律。1779年库仑发表了《论简单机械》——一篇分析机械间的摩擦的论文。战后退休的库仑协助新政府制定了一个重量和测量标准。库仑这个用于测量电荷的单位就是以其名字命名的。

Navier (1785-1836, as shown in Figure 6.11), was educated at the Ecole Polytechnique and became a professor there in 1831. Whereas the famous memoir of Coulomb (1773) contained the correct solution to numerous important problems in mechanics of materials, it took engineers more than forty years to understand them correctly and to use them in practical application.

纳维(1785—1836,如图 6.11 所示)曾就读于巴黎综合理工大学,并在 1831 年成为该校的一名教授。纪念库仑的专著(1773 年)提出了许多正确的计算方法用于解决材料力学中的许多重要问题。但足足花了超过四十年的时间,工程师们才能正确地理解这一理论并将其应用于工程实践。

Figure 6.11　Coulomb and Navier
图 6.11　库仑和纳维

In 1826 he published his Lecons (lecture notes) which is considered the first great textbook in mechanics for engineering. In it he developed the first general theory of elastic solids as well as the first systematic treatment of the theory of structures.

1826 年他发表了被认为是第一部伟大的工程力学方面的教科书。在书中他提出了第一个关于固体弹性的一般理论,这也是第一个系统地处理结构问题的理论。

It should be noted that no clear division existed between the theory of elasticity and the theory of structures until about the middle of the nineteenth century (Coulomb and Navier would today be considered professional structural engineers).

应当指出的是,直到 19 世纪中叶才将弹性力学和结构力学完全的区分开(今天库仑和纳维仍被认为是专业的结构工程师)。

Three other structural engineers who pioneered the development of the theory of elasticity from that point on were Lame, Clapeyron and de Saint-Venant. Lame published the first book on elasticity in 1852, and credited Clapeyron for the theorem of equality between external and internal work. De Saint-Venant was perhaps the greatest elasticians who according to Southwell "combined with high mathematical ability an essentially practical outlook which gave direction to all his work."

其他三位被认为是弹性理论先驱者的结构工程师分别是拉梅、克拉珀龙和圣维南。1852 年拉梅发表了第一部弹性力学著作,引用了克拉珀龙等效理论来研究物体外部和内部受力平衡的问题。圣维南也许是最伟大的弹性力学家,按照索思韦尔的说法,"结合运用超高的数学分析运算能力为其在工程中的应用指明了研究方向"。

# 6.8 The Modern Period (1857 to Present) 新世纪时期(1857 年至今)

## 1. Structures/Mechanics 结构力学

From 1857 the evolution of a comprehensive theory of structures proceeded at astonishing rate now that the basic and requisite principles had been determined.

自 1857 年以来，结构理论的全面发展过程是惊人的，基本和必要的原则已经确定。

Great contributors in that period include: Maxwell (first analysis of indeterminate structures), Culmann (graphics statics), Mohr (Mohr's circle, indeterminate analysis), Castigliano (1st and 2nd theorems), Cross (moment distribution), Southwell (relaxation method).

这一时期伟大的贡献包括：麦斯韦尔(静不定结构分析)，库尔曼(图形静力学)，摩尔(摩尔圆，不确定性分析)，卡氏(第一和第二定理)，截面(弯矩分布)，索思韦尔(松弛法)。

## 2. Eiffel Tower 埃菲尔铁塔

The Eiffel Tower was designed and built by the French civil engineer Alexandre Gustave Eiffel for the Paris World's Fair of 1889. The tower, with its modern broadcasting antennae, is 300 m (984ft) high. The lower section consists of four immense arched legs set on masonry piers. The legs curve inward until they unite in a single tapered tower. Platforms, each with an observation deck, are at three levels; on the first is also a restaurant.

埃菲尔铁塔是由法国的土木工程师亚历山大·古斯塔夫·埃菲尔为 1889 年巴黎世界博览会建造的。埃菲尔铁塔设有现代的广播天线，高 300 米(984 英尺)。下段是四个巨大的拱形钢腿放在砖石桥墩上，四个钢腿向内弯曲直到它们组成一个单一的锥形塔。有三个水平平台，每一个都有观光台；第一观光台也是一个餐厅。

## 3. Sullivan 沙利文

Sullivan(1856-1924) was an American architect, whose brilliant early designs for steel-frame skyscraper construction led to the emergence of the skyscraper as the distinctive American building type. Through his own work, especially his commercial structures, and as the founder of what is now known as the Chicago school of architects, he exerted an enormous influence on American architecture of $20^{th}$ century. His most famous pupil was the architect Frank Lloyd Wright, who acknowledged Sullivan as his master.

沙利文(1856—1924)是美国的建筑师，其辉煌的早期设计是钢框架的摩天大楼，他的设计导致摩天大楼的出现并成为独特的美国式建筑。通过自己的作品，尤其是他的商业化的建筑，并且作为芝加哥建筑学校的创始人，他对 20 世纪的美国式建筑风格产生了巨大的影响。沙利文最著名的学生是建筑师弗兰克·劳埃德·莱特，他把沙利文当成自己的导师。

After studying architecture at the Massachusetts Institute of Technology, he spent a year in Paris. Settling in Chicago in 1875, he was employed as a draftsman, then in 1881 formed a partnership with Dankmar Adler. Together they produced more than 100 buildings.

在麻省理工学院学习建筑之后，沙利文在巴黎待了一年。1875 年他作为一个绘图员在芝加哥定居，然后在 1881 年与丹克马尔·阿德勒成为合作伙伴。他们在一起建造了 100 多个建筑。

Adler secured the clients and handled the engineering and acoustical problems, while Sullivan concerned himself with the architectural designs. One of their earliest and most distinguished joint enterprises was the ten-story Auditorium Building (1886-1889) in Chicago. The Wainwright Building, also ten stories high, with a metal frame, was completed in 1891 in Saint Louis, Missouri. In 1895 the Sullivan-Adler partnership was dissolved, leading to a decline in Sullivan's practice. The Carson Pirie Scott (originally Schlesinger and Meyer) Department Store, Chicago, regarded by many as Sullivan's masterpiece, was completed in 1904.

阿德勒负责获得客户并处理工程和声学问题，而沙利文自己负责建筑设计。他们早期最著名的一个合作成果是在芝加哥的十层的礼堂大楼(1886—1889)。温赖特建筑也有十层楼高，是金属框架结构，于 1891 年在密苏里的圣路易斯完成。1895 年沙利文和阿德勒的合伙解散，导致沙利文的业务减少。完成于 1904 年并位于芝加哥的卡尔森·皮里·史葛(原名施莱辛格和迈耶)百货公司被认为是沙利文的杰作。

### 4. Khan 汗

Fazlur Khan was born in 1929 in Dacca India. After obtaining a B.A. in engineering from the University of Dacca in 1950, Khan worked as assistant engineer for the India Highway Department and taught at the University of Dacca. Qualifying for a scholarship in 1952, he enrolled at the University of Illinois, Champaign-Urbana, where he received master's degrees in both applied mechanics and structural engineering and a Ph.D. in structural engineering. He returned briefly to Pakistan and won an important position as executive engineer of the Karachi Development Authority. Frustrated by administrative demands that kept him from design work, however, he returned to the United States and joined the prestigious architectural firm of Skidmore, Owings & Merrill in Chicago in 1955, eventually becoming a partner (1966).

法兹勒·汗 1929 年出生于印度达卡。1950 从达卡大学工程获得学士学位之后，汗成为印度公路部的助理工程师并在达卡大学教学。1952 年申请到奖学金后，他被伊利诺伊大学香槟分校录取，在那里他获得了应用力学和结构工程硕士学位以及结构工程学博士。他短暂返回巴基斯坦并获得一个重要的职位，成为卡拉奇发展局的执行工程师。然而，从事行政工作而远离设计工作使他很沮丧，他又回到美国并在 1955 年于芝加哥加入了奥因斯和美林建筑协会下著名的斯基德莫尔建筑公司，最终在 1966 年成为合作人。

Among his many designs for skyscrapers are Chicago's John Hancock Center (1970) and the Sears Tower (1973), which are among the world's tallest buildings, and One Shell Plaza in Houston, Texas. The Sears Tower was his first skyscraper to employ his "bundled tube" structural system. The system was innovative because it minimized the amount of steel needed for high towers, eliminated the need for internal wind bracing (since the perimeter columns carried the wind loadings), and permitted freer organization of the interior space.

他设计的许多摩天大楼中有芝加哥的约翰·汉考克中心(1970)和西尔斯大厦(1973)，它们是世界上最高的建筑物之一，还有位于得克萨斯州休斯敦市的贝壳广场大厦。西尔斯大厦是他第一次利用其"束筒"结构体系建造的摩天大楼。该系统的创新之处在于它使高塔的用钢量最小化，消除了内部斜撑的设置(因为边缘的柱子才能抵抗风荷载)，并允许内部空间的自由设置。

His later projects included the strikingly different Haj Terminal of the King Abdul Aziz International Airport, Jiddah, Saudi Arabia (1976-1981), and King Abdul Aziz University, also in Jiddah (1977-1978).

他后来的项目包括标新立异的阿卜杜勒·阿齐兹国王国际机场(1976—1981)和也在吉达的阿卜杜勒·阿齐兹国王大学(1977—1978)。

## New Words and Expressions　生词和短语

1. evolution　　*n.* 演变；进化论；进展
2. antiquity　　*n.* 高龄；古物；古代的遗物
3. pyramid　　*n.* 金字塔；角锥体
4. cathedral　　*n.* 大教堂
5. philosopher　　*n.* 哲学家；哲人
6. collapse　　*n.* 倒塌；失败；衰竭
　　　　　　*vt.* 使倒塌，使崩溃；使萎陷
7. obscure　　*adj.* 模糊的；晦涩的；昏暗的
8. conqueror　　*n.* 征服者；胜利者
9. Pantheon　　*n.* 万神殿；名流群
10. cylinder　　*n.* 圆筒；汽缸；[数] 柱面；圆柱状物
11. pier　　*n.* 码头，直码头；桥墩
12. dome　　*n.* 圆屋顶
13. porch　　*n.* 门廊；走廊
14. artillery engineer　　炮兵工程师
15. sanitation　　*n.* 环境卫生；卫生设备；下水道设施
16. practical hydraulics　　实用水力学
17. thesaurus　　*n.* 宝库；辞典；知识宝库
18. The Medieval Period　　中世纪时期
19. algebra　　*n.* 代数学
20. Gothic period　　哥特式时期
21. barrel vault　　筒形穹顶
22. thrust　　*n.* [力] 推力；刺
　　　　　*vt.* 插；插入；推挤
23. innovation　　*n.* 创新，革新；新方法
24. discrete　　*adj.* 离散的，不连续的
25. continuous　　*adj.* 连续的，持续的

26. buttress    *n.* 扶壁；拱壁；支撑物
27. Renaissance    *n.* 文艺复兴
28. manuscript    *n.* 手稿；原稿
29. ensile strength    抗拉强度
30. cantilevered beam    悬臂梁
31. architectural functionalism    建筑功能主义
32. framework    *n.* 框架，骨架；结构，构架
33. elasticity    *n.* 弹性；弹力；灵活性
34. planetary motion    行星运动
35. orbital motion    轨道运动，轨道运行
36. friction    *n.* 摩擦，[力] 摩擦力
37. draftsman    *n.* [测] 绘图员；起草者
38. acoustical    *adj.* [声] 声学的；听觉的；音响的
39. bundled tube    束筒

# Exercises  练习

Ⅰ. Write a T in front of a statement if it is true according to the text and write an F if it is false.

1. The Greek philosopher Pythagoras founded his famous school, which was primarily a secret religious society, at Crotona in southern Italy.

2. The greatest of the Greeks was Archimedes who was one of the greatest physicist of the ancient world and one of its greatest mathematician.

3. Science made much more progress under the Romans than under the Greeks.

4. Since the combination of ribs and piers relieved the intervening vertical wall spaces of their supportive function, these walls could be built thinner and could even be opened up with large windows or other glazing.

5. During the Renaissance there was a major revival of interest in music and art.

6. Hooker was not only very famous in the study of rational mechanics, but also contributed to many other fields of science.

Ⅱ. Complete the following sentences.

1. _____ was the builder of the step pyramid of Sakkara about 3000 B.C., and yielded great influence over ancient Egypt.

2. The Medieval Period, also called _____, was marked by _____.

3. Hooker was not only very famous in the study of _____, but also contributed to many other fields of science.

4. _____established the modern science of dynamics by formulating his three laws of

motion.

5. In 1777, Coulomb invented the _____ for measuring the force of magnetic and electrical attraction.

6. The Eiffel Tower was designed and built by the French civil engineer _____.

Ⅲ. Translate the following sentences into Chinese.

1. If I have been able to see a little farther than some others, it was because I stood on the shoulders of giants.

2. The builders were guided by rules of experiences, which were passed from generation to generation, and seldom supplemented by new knowledge.

3. During that period, the Arabs carried the torch of knowledge, gave birth to algebra, translated some of the great books of the Library of Alexandria.

4. His influence on his contemporaries and immediate followers was very strong and has been felt even in the 20th century.

5. During the following two and a half years, Newton established the modern science of dynamics by formulating his three laws of motion.

6. It should be noted that no clear division existed between the theory of elasticity and the theory of structures until about the middle of the nineteenth century.

Ⅳ. Translate the following sentences into English.

1. 在他的学校听讲座时他既不需要带课本也不需要做笔记。

2. 这些卷轴中的许多后来通过阿拉伯人的翻译引起了西方世界的关注。

3. 它的巨大的规模和大胆的技术创新，使其成为世界上主要古迹之一。

4. 胡克不仅在弹性力学的研究方面非常有名，而且对许多其他的科学领域也做出了贡献。

5. 除了科学，牛顿在炼金术、神秘主义和神学方面也表现出了兴趣。

6. 他非常尊重在建立建筑学的理论方面发挥了巨大作用的自然世界。

# Answers　答案

Ⅰ.

1. T　2. T　3. F　4. T　5. F　6. F

Ⅱ.

1. Imhotep

2. the Dark Ages, a general decline of civilization

3. elasticity

4. Newton

5. torsion balance

6. Alexandre Gustave

Ⅲ.

1. 如果说我能看得更远一些，那是因为我站在巨人的肩膀上。

2. 建设者遵循经验规则，并世代相传，而且很少补充新的知识。

3. 在这期间，阿拉伯人带着知识的火炬，创造了代数，翻译了一些亚历山大图书馆的伟大书籍。

4. 他对他的同时代人和直接追随者产生了非常强大的影响，甚至在 20 世纪也是如此。

5. 在接下来的两年半的时间里，牛顿通过形成三大运动定律，建立了现代科学的动力学。

6. 应该注意的是，在弹性理论和结构理论之间没有明确的区分，直到大约 19 世纪中期。

Ⅳ.

1. At his school he allowed neither textbooks nor recording of notes in lectures.

2. Many of these scrolls were subsequently brought to the attention of the western world through translations by the Arabs.

3. Its huge size and daring technical innovations make it one of the world's key monuments.

4. Hooker was not only very famous in the study of elasticity, but also contributed to many other fields of science.

5. In addition to science, Newton also showed an interest in alchemy, mysticism, and theology.

6. He also had tremendous respect for the natural world which played an enormous role in forging his theories about architecture.

# Chapter 7　Reinforced Concrete Architecture
# 第 7 章　钢筋混凝土

Reinforced concrete is concrete in which reinforcement bars ("rebars"), reinforcement grids, plates or fibers have been incorporated to strengthen the concrete in tension. It was invented by French gardener Joseph Monier in 1849 and patented in 1867. The term Ferro Concrete refers only to concrete that is reinforced with iron or steel. Other materials used to reinforce concrete can be organic and inorganic fibers as well as composites in different forms. Concrete is strong in compression, but weak in tension, thus adding reinforcement increases the strength in tension. In addition, the failure strain of concrete in tension is so low that the reinforcement has to hold the cracked sections together. For a strong, ductile and durable construction the reinforcement shall have the following properties:
- High strength;
- High tensile strain;
- Good bond to the concrete;
- Thermal compatibility;
- Durability in the concrete environment.

钢筋混凝土就是混凝土中应用钢筋、钢筋网、钢板、纤维共同作用来承受拉力。1849年，一位名叫约瑟夫·莫尼尔的法国园丁发明了钢筋混凝土，并于1867年申请专利。术语中所指的钢筋混凝土仅指利用铸铁或者钢进行增强的混凝土。此外一些有机纤维、无机纤维或者很多形式的复合材料也可用来增强混凝土。混凝土的抗压能力很强，但是抗拉能力很弱，因此这些增强措施提高了混凝土的抗拉能力。另外因为混凝土的抗拉能力很弱，所采用的增强措施必须具备将混凝土破碎的部分连接起来的功能。一个结构坚固、延展性好、耐久的混凝土结构需要具备以下的性能：
- 强度高；
- 抗拉刚度高；
- 混凝土之间黏结性能好；
- 温度相容性；
- 在混凝土环境中耐久性好。

In most cases reinforced concrete uses steel rebars that have been inserted to add strength.
在大多数情况下，利用在混凝土中配置钢筋的方式来提高强度。

# 7.1 Use in Construction  在建筑中的应用

Concrete is reinforced to give it extra tensile strength; without reinforcement, many concrete buildings would not have been possible.

混凝土的增强钢筋主要是提高混凝土的抗拉强度；正是混凝土采用钢筋才使得许多混凝土建筑成为可能。

Reinforced concrete can encompass many types of structures and components, including slabs, walls, beams, columns, foundations, frames and more.

钢筋混凝土包括许多种类的结构和构件，如楼板、墙、梁、柱、基础、框架等。

Reinforced concrete can be classified as precast or cast in-situ concrete.

钢筋混凝土可以归类为预制混凝土和现浇混凝土。

Much of the focus on reinforcing concrete is placed on floor systems. Designing and implementing the most efficient floor system is key to creating optimal building structures. Small changes in the design of a floor system can have significant impact on material costs, construction schedule, ultimate strength, operating costs, occupancy levels and end use of a building.

工程中主要承担荷载的钢筋混凝土构件是楼盖系统。设计并施工建造最有效的楼盖系统是达到理想的建筑结构的关键。楼盖设计的一个很小的改变都会对材料费、施工进度、强度极限、生产费用、居住水平和建筑的最终用途产生很大的影响。

# 7.2 Behavior of Reinforced Concrete  钢筋混凝土的性能

### 1. Materials  材料

Concrete is a mixture of coarse (stone or brick chips) and fine (generally sand) aggregates with a binder material (usually Portland cement). When mixed with a small amount of water, the cement hydrates form microscopic opaque crystal lattices encapsulating and locking the aggregate into a rigid structure. Typical concrete mixes have high resistance to compressive stresses (about 4,000psi (28MPa)); however, any appreciable tension (e.g. due to bending) will break the microscopic rigid lattice, resulting in cracking and separation of the concrete. For this reason, typical non-reinforced concrete must be well supported to prevent the development of tension.

混凝土由是粗的(石头或碎砖块)和细的(普通的沙子)骨料与胶凝材料(通常是硅酸盐水泥)混合制成的。当掺入少量水后，水泥与水化合形成微小的不透明的晶体，将整体封闭成一个稳定的结构。混凝土混合物的典型特征是有很强的受压能力(大约是 4000psi(28MPa))；然而，任何拉力(如由于拉力造成的弯曲)都会打破微观的晶格从而导致混凝土出现裂缝和分离。正是由于这个原因，无钢筋的素混凝土主要是防止受拉破坏。

If a material with high strength in tension, such as steel, is placed in concrete, then the composite material, reinforced concrete, resists not only compression but also bending and other direct tensile actions. A reinforced concrete section where the concrete resists the compression and steel resists the tension can be made into almost any shape and size for the construction industry.

像复合材料、钢筋混凝土等有很强的抗拉能力的材料，不仅能够抗压抗拉还能够抗弯或者抵抗其他形式的拉力。钢筋和混凝土分工协作，混凝土抗压，钢筋抗拉，在建筑行业中钢筋混凝土可以被制成任何形状和任何大小的形式。

### 2. Key Characteristics　主要性能

Three physical characteristics give reinforced concrete its special properties.

有三个使钢筋混凝土有独特性能的物理特征。

First, the coefficient of thermal expansion of concrete is similar to that of steel, eliminating large internal stresses due to differences in thermal expansion or contraction.

第一，混凝土的温度膨胀系数与钢筋相近，消除了由于温度膨胀系数不同导致的过大的应力和收缩。

Second, when the cement paste within the concrete hardens this conforms to the surface details of the steel, permitting any stress to be transmitted efficiently between the different materials. Usually steel bars are roughened or corrugated to further improve the bond or cohesion between the concrete and steel.

第二，水泥浆依附在钢筋表面，依据钢筋的螺纹形状而硬化，使得任何形式的应力能够在不同材料间高效地传递。通常钢筋是粗糙的或者有螺纹的，这样可以进一步增大混凝土和钢筋之间的黏结力。

Third, the alkaline chemical environment provided by the alkali reserve (KOH, NaOH) and the portlandite (calcium hydroxide) contained in the hardened cement paste causes a passivating film to form on the surface of the steel, making it much more resistant to corrosion than it would be in neutral or acidic conditions. When the cement paste exposed to the air and meteoric water reacts with the atmospheric $CO_2$, portlandite and the Calcium Silicate Hydrate (CSH) of the hardened cement paste become progressively carbonated and the high pH gradually decreases from 13.5 – 12.5 to 8.5, the pH of water in equilibrium with calcite (calcium carbonate) and the steel is no longer passivated.

第三，存留的碱性物质(KOH，NaOH)和水泥浆中含有的氢氧钙石形成了碱性的化学环境，无形中在钢筋的表面形成了一层钝化膜，使得钢筋比在中性环境和酸性环境中更加抗腐蚀。当水泥浆被暴露在空气中时，大气水和空气中的 $CO_2$ 相互作用使得硬化的水泥浆中的氢氧钙石和水化硅酸钙含碳量逐渐增加，pH 值逐渐从 13.5～12.5 减少到 8.5，水和碳酸钙维持的恒定 pH 值开始变化，钢筋也不再是钝化的。

As a rule of thumb, only to give an idea on orders of magnitude, steel is protected at pH above 11 but starts to corrode below 10 depending on steel characteristics and local physico-chemical conditions when concrete becomes carbonated. Carbonation of concrete along with chloride ingress are amongst the chief reasons for the failure of reinforcement bars in concrete.

一般说来，当混凝土碳化时，根据钢筋的性能和当地的物理化学条件，钢筋在 pH 大于

11 时处于保护状态，但 pH 小于 10 时开始腐蚀。碳化混凝土被氯离子侵蚀是混凝土中的钢筋被腐蚀的最主要原因。

The relative cross-sectional area of steel required for typical reinforced concrete is usually quite small and varies from 1% for most beams and slabs to 6% for some columns. Reinforcing bars are normally round in cross-section and vary in diameter. Reinforced concrete structures sometimes have provisions such as ventilated hollow cores to control their moisture & humidity.

相对来说，一般的钢筋混凝土对钢筋截面面积的要求非常小，通常只有大多数的梁、板配筋率要求为 1%，有的柱配筋率要求为 6%。钢筋通常是直径不同的圆钢。为了控制合理的湿度，钢筋混凝土结构有时会对通风口有要求。

Distribution of concrete (in spite of reinforcement) strength characteristics along the cross-section of vertical reinforced concrete elements is inhomogeneous according to article *Concrete Inhomogeneity of Vertical Cast-In-Situ Elements In Frame-Type Buildings*.

根据文章《框架结构中竖向现浇构件混凝土的不均匀性》我们可以了解到，竖向钢筋混凝土构件虽然得到了增强，但在内部沿着钢筋，混凝土的应力分布也是不均匀的。

## 7.3　Reinforcement and Terminology of Beams
　　　梁的增强和概念

A beam bends under bending moment, resulting in a small curvature. At the outer face (tensile face) of the curvature the concrete experiences tensile stress, while at the inner face (compressive face) it experiences compressive stress.

梁在弯矩的作用下会产生一个小的弯曲变形。在弯曲外表面(拉伸面)混凝土产生拉应力，在弯曲内表面(压缩面)混凝土产生压应力。

A singly-reinforced beam is one in which the concrete element is only reinforced near the tensile face and the reinforcement, called tension steel, is designed to resist the tension.

单筋梁是指只在混凝土构件受拉区设置抗拉钢筋，用来抵抗拉力。

A doubly-reinforced beam is one in which besides the tensile reinforcement the concrete element is also reinforced near the compressive face to help the concrete resist compression. The latter reinforcement is called compression steel. When the compression zone of a concrete is inadequate to resist the compressive moment (positive moment), extra reinforcement has to be provided if the architect limits the dimensions of the section.

双筋梁是指混凝土梁既设置抗拉钢筋来抵抗拉力，也在受压区设置抗压钢筋。当构件受压区混凝土不足以抵抗弯矩(正弯矩)所产生的压力时，且构件的设计截面尺寸受到了限制，那就需要额外的钢筋增强。

An under-reinforced beam is one in which the tension capacity of the tensile reinforcement is smaller than the combined compression capacity of the concrete and the compression steel (under-reinforced at tensile face). When the reinforced concrete element is subject to increasing bending moment, the tension steel yields while the concrete does not reach its ultimate failure condition. As the tension steel yields and stretches, an "under-reinforced" concrete also yields in a ductile manner, exhibiting a large deformation and warning before its ultimate failure. In this case the yield stress of the steel governs the design.

适筋梁是梁受拉区的抗拉能力比受压区的混凝土和抗压钢筋的抗压能力弱的梁(受拉区适筋)。当混凝土构件的弯矩逐渐增加,抗拉钢筋屈服,此时混凝土并没有达到极限破坏条件。随着抗拉钢筋的屈服和拉伸,"适筋"混凝土也以一种韧性破坏的方式屈服,并在最终破坏前有较大形变,产生预兆。所以,在设计时应该主要考虑钢筋的屈服强度。

An over-reinforced beam is one in which the tension capacity of the tension steel is greater than the combined compression capacity of the concrete and the compression steel (over-reinforced at tensile face). So the "over-reinforced concrete" beam fails by crushing of the compressive-zone concrete and before the tension zone steel yields, which does not provide any warning before failure as the failure is instantaneous.

超筋梁是抗拉钢筋的抗拉能力比混凝土和抗压钢筋的抗压能力强的梁 (在受拉区超筋)。在受拉区钢筋屈服前,超筋梁受压区的混凝土就发生破坏,破坏非常突然,没有任何预兆。

A balanced-reinforced beam is one in which both the compressive and tensile zones reach yielding at the same imposed load on the beam, and the concrete will crush and the tensile steel will yield at the same time. This design criterion is however as risky as over-reinforced concrete, because failure is sudden as the concrete crushes at the same time of the tensile steel yields, which gives a very little warning of distress in tension failure.

少筋梁是指受拉区和受压区在同一荷载作用下达到屈服强度,混凝土突然破坏,抗拉钢筋也同时破坏。然而这种设计标准非常危险,混凝土和抗拉钢筋同时破坏,在破坏前几乎没有任何预兆。

Steel-reinforced concrete moment-carrying elements should normally be designed to be under-reinforced so that users of the structure will receive warning of impending collapse.

重要的钢筋混凝土构件通常都设计成适筋梁形式,这样结构使用者在其即将破坏前会收到预兆。

The characteristic strength is the strength of a material where less than 5% of the specimen shows lower strength.

强度是材料的力学性能,通常取值低于材料的试件试验强度的5%。

The design strength or nominal strength is the strength of a material, including a material-safety factor. The value of the safety factor generally ranges from 0.75 to 0.85 in Allowable Stress Design.

材料强度包括材料强度的计值或标准值以及安全系数。在许用应力设计中,安全系数的值通常为 0.75~0.85。

The ultimate limit state is the theoretical failure point with a certain probability. It is stated under factored loads and factored resistances.

极限状态是理论上的破坏点。它考虑到了荷载和抗力的因素。

## 7.4 Common Failure Modes of Steel Reinforced Concrete 钢筋混凝土常见的破坏形式

Reinforced concrete can fail due to inadequate strength, leading to mechanical failure, or due to a reduction in its durability. Corrosion and freeze/thaw cycles may damage poorly designed or constructed reinforced concrete. When rebar corrodes, the oxidation products (rust) expands and tends to flake, cracking the concrete and unbonding the rebar from the concrete. Typical mechanisms leading to durability problems are discussed below.

钢筋混凝土由于强度不足和耐久性差会发生机械破坏。腐蚀、冻融都可能会导致设计有缺陷或者建好的钢筋混凝土结构破坏。当钢筋受到侵蚀时，氧化物(锈)的面积增加并开始剥落，导致混凝土开裂并使钢筋与混凝土剥离。下面将讨论典型的机械破坏所导致的耐久性问题。

Mechanical failure: Cracking of the concrete section can not be prevented; however, the size of and location of the cracks can be limited and controlled by reinforcement, placement of control joints, the curing methodology and the mix design of the concrete. Cracking defects can allow moisture to penetrate and corrode the reinforcement. This is a serviceability failure in limit state design. Cracking is normally the result of an inadequate quantity of rebar, or rebar spaced at too great a distance. The concrete then cracks either under excess loading, or due to internal effects such as early thermal shrinkage when it cures.

机械破坏：混凝土截面开裂是不能避免的，但是可以通过技术手段来控制裂缝的大小和位置，如配筋、设置伸缩缝、养护方法或混凝土的配合比设计。裂缝会导致水汽侵入并侵蚀钢筋。这就是极限状态设计中的正常使用极限状态破坏。产生裂缝通常是由于配筋不足、钢筋间距过大造成的。混凝土在荷载过大时会产生裂缝，内部效应如混凝土硬化时热收缩也会产生裂缝。

Ultimate failure leading to collapse can be caused by crushing of the concrete, when compressive stresses exceed its strength; by yielding or failure of the rebar, when bending or shear stresses exceed the strength of the reinforcement; or by bond failure between the concrete and the rebar.

导致混凝土最终破坏的原因可能有以下几种：压力超过抗压极限；钢筋屈服或破坏；弯曲应力或剪切应力超过钢筋混凝土承载极限；钢筋和混凝土之间的黏结失效破坏。

## 7.5 Carbonation 碳化作用

Carbonation, or neutralisation, is a chemical reaction between carbon dioxide in the air with calcium hydroxide and hydrated calcium silicate in the concrete. The water in the pores of Portland cement concrete is normally alkaline with a pH in the range of 12.5 to 13.5. This highly alkaline environment is one in which the embedded steel is passivated and is protected from corrosion. According to the Pourbaix diagram for iron, the metal is passive when the pH is above

9.5. The carbon dioxide in the air reacts with the alkali in the cement and makes the pore water more acidic, thus lowering the pH. Carbon dioxide will start to carbonate the cement in the concrete from the moment the object is made. This carbonation process will start at the surface, then slowly move deeper and deeper into the concrete. The rate of carbonation is dependent on the relative humidity of the concrete — a 50% relative humidity being optimal. If the object is cracked, the carbon dioxide in the air will be better able to penetrate into the concrete. When designing a concrete structure, it is normal to state the concrete cover for the rebar (the depth within the object that the rebar will be). The minimum concrete cover is normally regulated by design or building codes. If the reinforcement is too close to the surface, early failure due to corrosion may occur. The concrete cover depth can be measured with a cover meter. However, carbonated concrete only becomes a durability problem when there is also sufficient moisture and oxygen to cause electro-potential corrosion of the reinforcing steel.

碳化作用和中和作用是空气中的二氧化碳和混凝土中的氢氧化钙、含水硅酸钙之间的化学反应。硅酸盐混凝土气孔中的水分通常呈碱性，pH 值为 12.5～13.5。这种高碱性环境会使埋置的钢筋钝化，防止钢筋腐蚀。根据铁的 $\varphi$-pH 图，在 pH 值高于 9.5 时，金属是钝化的。空气中的二氧化碳与水泥中的碱性物质作用，会降低气孔中水的 pH 值。一旦有条件，二氧化碳会充满水泥的气孔。这种碳化过程从混凝土的表面开始，然后慢慢地深入混凝土中。碳化速率依赖于混凝土的相对湿度，相对湿度 50%是最佳条件。一旦混凝土某处产生破裂，空气中的二氧化碳就能顺利地进入混凝土内部。设计混凝土结构时，通常都会设计混凝土保护层来保护钢筋(整个构件内部的钢筋)。设计或施工规范都规定了最小的混凝土保护层厚度。如果钢筋距离外表面过近，钢筋可能会很快被腐蚀。混凝土保护层厚度可以用保护层厚度测定仪进行测量。然而，当水分和氧气含量很高时，钢筋会发生电离腐蚀，混凝土碳化就成了一个耐久性问题了。

One method of testing a structure for carbonation is to drill a fresh hole in the surface and then treat the cut surface with phenolphthalein indicator solution. This solution will turn pink when in contact with alkaline concrete, making it possible to see the depth of carbonation. An existing hole is no good because the exposed surface will already be carbonated.

检测一个结构的碳化程度的方法是在表面钻一个小洞，在小洞的切面用酚酞指示剂检测。弱指示剂显示粉色，则说明下层的混凝土呈碱性。钻出的小洞对混凝土是不利的，暴露在空气中的部分将很快被碳化。

## 7.6　Chlorides　氯化物

Chlorides, including sodium chloride, can promote the corrosion of embedded steel rebar if present in sufficiently high concentration. Chloride anions induce both localized corrosion (pitting corrosion) and generalized corrosion of steel reinforcements. For this reason, one should only use fresh raw water or potable water for mixing concrete, insure that the coarse and fine aggregates do not contain chlorides, and not use admixtures that contain chlorides.

高浓度的氯化物包括氯化钠会加快腐蚀配置的钢筋。氯离子会导致局部腐蚀(点状腐蚀)

和钢筋的腐蚀。因此，混合混凝土需要用未加工的水或者是饮用水，并确保粗骨料细骨料不含有氯离子，也不要加入带有氯离子的外加剂。

It was once common for calcium chloride to be used as an admixture to promote rapid set-up of the concrete. It was also mistakenly believed that it would prevent freezing. However, this practice has fallen into disfavor once the deleterious effects of chlorides became known. It should be avoided when ever possible.

氯化钙作为一种外加剂，以前常被用作加速混凝土的初凝，也经常被误认为能够抗冻。然而，当发现氯化物的腐蚀作用时这种观念被推翻了。要尽一切可能阻止氯化物的存在。

The use of de-icing salts on roadways, used to reduce the freezing point of water, is probably one of the primary causes of premature failure of reinforced or prestressed concrete bridge decks, roadways, and parking garages. The use of epoxy-coated reinforcing bars and the application of cathodic protection has mitigated this problem to some extent. Also FRP rebars are known to be less susceptible to chlorides. Properly designed concrete mixtures that have been allowed to cure properly are effectively impervious to the effects of deicers.

除冰盐通常被用在路面上，因为除冰盐可以降低水的凝固点，这也是钢筋混凝土和预应力混凝土桥面、道路、停车场过早被破坏的主要原因之一。使用环氧树脂涂覆钢、阴极保护能够在一定程度上减轻这种现象。另外 FRP 钢筋受氯离子的影响很小。设计合理的混凝土拌合物也可以有效地防止氯离子的侵蚀。

Another important source of chloride ions is from sea water. Sea water contains by weight approximately 3.5wt% salts. These salts include sodium chloride, magnesium sulfate, calcium sulfate, and bicarbonates. In water these salts dissociate in free ions ($Na^+$, $Mg^{2+}$, $Cl^-$, $SO_4^{2-}$, $HCO_3^-$) and migrate with the water into the capillaries of the concrete. Chloride ions are particularly aggressive for the corrosion of the carbon steel reinforcement bars and make up about 50% of these ions.

氯离子的另一个主要来源是海水。海水重量的 3.5 %是盐。这些盐包含氯化钠、硫酸镁、硫酸钙和碳酸氢盐。在水中，这些盐分呈游离状态，以 $Na^+$、$Mg^{2+}$、$Cl^-$、$SO_4^{2-}$、$HCO_3^-$ 离子形式存在，随着水进入混凝土内部。氯离子占这些离子总数的 50%，对碳化钢筋的腐蚀效果非常强。

In the 1960's and 1970's it was also relatively common for Magnesite, a chloride rich carbonate mineral, to be used as a floor-topping material. This was done principally as a levelling and sound attenuating layer. However it is now known that when these materials came into contact with moisture it produced a weak solution of hydrochloric acid due to the presence of chlorides in the magnesite. Over a period of time (typically decades) the solution caused corrosion of the embedded steel rebars. This was most commonly found in wet areas or areas repeatedly exposed to moisture.

20 世纪六七十年代比较常见的菱镁矿是一种含丰富氯离子的碳酸盐矿物，通常作为楼板面层材料，起到抄平和隔音的作用。然而现在才知道，这些材料与水接触后，氯离子会发生弱解反应产生盐酸。很长一段时间里(大概几十年)，这都是引起钢筋腐蚀的原因。在潮湿的地区和经常接触水分的地区，氯离子腐蚀的现象更为常见。

## New Words and Expressions　生词和短语

1. concrete　　*n.* 混凝土
2. cement　　*n.* 水泥
3. aggregate　　*n.* 骨料
4. mortar　　*n.* 砂浆
5. extender　　*n.* 掺合料
6. rigidity　　*n.* 刚度
7. creep　　*n.* 徐变
8. shear　　*n.* 剪力
9. tension　　*n.* 拉力
10. displacement　　*n.* 位移
11. stress　　*n.* 压力
12. strain　　*n.* 应变
13. slump　　*n.* 坍落度
14. stadium　　*n.* 龄期
15. reinforcing steel bar　　钢筋
16. reinforced concrete　　钢筋混凝土
17. reinforced concrete structure　　钢筋混凝土结构
18. prestressed reinforced concrete　　预应力钢筋混凝土
19. prestressed reinforcement　　预应力钢筋
20. cast-in-place reinforced concrete　　现浇钢筋混凝土结构
21. non-reinforced concrete　　素混凝土
22. cover to reinforcement　　钢筋保护层
23. class of cube strength　　强度等级
24. mixing time　　拌和时间
25. concrete transportation time　　混凝土运输时间
26. concreting temperature　　浇筑温度
27. mixing proportion　　配合比
28. concrete vibrating　　振捣
29. cover to reinforcement　　钢筋保护层
30. shear wall　　剪力墙
31. shear deformation　　剪切变形
32. percentage of elongation　　延伸率
33. steel wire　　箍筋
34. compressive strength　　抗压强度
35. bending strength　　抗弯强度
36. torsional strength　　抗扭强度

37. tensile strength　　抗拉强度
38. crack　　*n.* 裂缝
39. yield　　*n.* 屈服
40. yield point　　屈服点
41. yield load　　屈服荷载
42. limit of yielding　　屈服极限
43. yield strength　　屈服强度
44. reinforcement ratio　　配筋率
45. fatigue strength　　疲劳强度
46. frame structure　　框架结构
47. pouring　　*v.* 浇注
48. concreting　　*n.* 浇注混凝土
49. rigid frame　　刚架
50. brittle failure　　脆性破坏

# Exercises　练习

Ⅰ. Write a T in front of a statement if it is true according to the text and write an F if it is false.

1. Smaller particles up to 1/4 in size are called fine aggregates.

2. When concrete is poured, the free water need for the hydration process evaporates over a period of time and the concrete will sink.

3. In normall concrete, the reinforcing is protected by naturally high alkalinity of the concrete with a pH of about 10.

4. The concrete testing programme should be designed to yield the necessary information to properly assess the structure.

5. When selecting repair materials, a primary objective is to match the properties of the repair material as closely as possible with the parent concrete.

6. An effective method to control the corrosion of steel in contaminated concrete is cathodic protection.

Ⅱ. Complete the following sentences.

1. Reinforced concrete was invented by _____.

2. For a strong, ductile and durable construction the reinforcement shall have the following properties: _____, _____, _____, _____, _____, _____ .

3. Concrete is strong in compression, but _____, thus adding reinforcement _____ .

4. Reinforced concrete can encompass many types of structures and components, including _____, _____, _____, _____, _____, _____ and more.

5. Concrete is a mixture of_____.

Ⅲ. Translate the following sentences into Chinese.

1. Concrete is strong in compression, but weak in tension.

2. Concrete is reinforced to give it extra tensile strength.

3. A beam bends under bending moment, resulting in a small curvature.

4. The characteristic strength is the strength of a material where less than 5% of the specimen shows lower strength.

5. The water in the pores of Portland cement concrete is normally alkaline with a pH in the range of 12.5 to 13.5.

Ⅳ. Translate the following sentences into English.

1. 在大多数情况下，利用在混凝土中配置钢筋的方式来提高强度。
2. 单筋梁是指只在混凝土构件受拉区设置抗拉钢筋，用来抵抗拉力。
3. 材料强度包括材料强度的设计值以及安全系数。
4. 极限状态是理论上的破坏点。
5. 氯离子另一个主要来源是海水。

## Answers　答案

Ⅰ.

1. T　2. F　3. F　4. T　5. T　6. T

Ⅱ.

1. French gardener Joseph Monier in 1849 and patented in 1867

2. high strength, high tensile strain, good bond to the concrete, thermal compatibility, durability in the concrete environment

3. weak in tension, increases the strength in tension

4. slabs, walls, beams, columns, foundations, frames

5. coarse and fine aggregates with a binder material

Ⅲ.

1. 混凝土的抗压能力很强，但是抗拉能力很弱。
2. 混凝土的增强钢筋主要是提高混凝土的抗拉强度。
3. 梁在弯矩作用下会产生一个小的弯曲变形。
4. 强度是材料的力学性能，通常取值低于材料的试件试验强度的5%。

5. 硅酸盐混凝土气孔中的水分通常呈碱性，pH 值在 12.5～13.5 之间。

Ⅳ.

1. In most cases reinforced concrete uses stell rebars that have been inserted to add strength.

2. A singly-reinforced beam is one in which the concrete is only reinforced near the tensile face and the reinforcement, called tension stell, is designed to resist the tension.

3. The design strength or nominal strength is the strength of a material, including a material-safety factor.

4. The ultimate limit state is the theoretical failure point with a certain probability.

5. Another important source of ions is from sea water.

# Chapter 8　Steel Structure
# 第 8 章　钢　结　构

## 8.1　History of Steel Structures　钢结构的发展历史

Steel structure refers to a broad range of building construction in which steel plays the leading role. Most steel construction consists of large-scale buildings or engineering works, with the steel generally in the form of beams, girders, bars, plates, and other members shaped through the hot-rolled process.

钢结构应用的范围很广，通常指的是以钢材作为主要材料的房屋建筑。钢结构大多数用在大型建筑或工程中。一般通过热轧塑造而成，形成次梁、主梁、条状、平面或其他形状的钢材。

Steel is an alloy of iron-carbon, the history of steel structure application is closely related to the development of iron-making and steel-making technology. As early as around 2000 BC, the home of numerous early civilizations, Mesopotamian plain (An ancient region of southwest Asia between the Tigris and Euphrates rivers in modern-day Iraq), appeared in the early iron-making technology.

钢是一种铁碳合金，钢结构应用的历史与炼铁和炼钢的技术发展紧密相关，早在公元前 2000 年左右，作为很多早期文明家园的美索不达米亚平原(亚洲西南部的一个古代地区，位于现代伊拉克的底格里斯河和幼发拉底河之间)就出现了早期的炼铁技术。

During the Warring States period of China, the iron-making technology has been very popular. About A.D. 65, the material wrought iron has been successfully used in the world's first iron chain suspension bridge — Lan Jin Bridge (as shown in Figure 8.1). Since then, dozens of chains bridges have been built. The largest span is DaDuhe Bridge in Lu Ding County which was built in 1705 (as shown in Figure 8.2), which was 74 years earlier than the England Bridge (as shown in Figure 8.3) known as the world's first cast iron arch bridge. The above shows that ancient Chinese used to occupy the world's leading position in the application of the iron structure.

战国时期的中国，炼铁技术已经非常流行，大约在公元 65 年，锻铁材料已成功地应用于世界上第一个铁链吊桥——兰津桥(如图 8.1 所示，桥名霁虹，位于西汉的兰津古渡，在云南大理永平县与保山市交界处的澜沧江上)。从那时起，拥有数十根链条的桥开始建造。最大跨度的链锁吊桥是建于 1705 年的位于泸定县的泸定桥(如图 8.2 所示)，比被称为世界第一的铸铁拱桥英格兰桥(如图 8.3 所示)早出现 74 年。以上表明古代中国在铁结构的应用上占据世界领先地位。

Figure 8.1　Lan Jin Bridge(A.D. 65)
图 8.1　兰津桥(公元 65 年)

Figure 8.2　Lu Ding Bridge (A.D. 1705)
图 8.2　泸定桥(公元 1705 年)

Figure 8.3　The Cast Iron Bridge in England
图 8.3　英格兰铸铁桥

In foreign countries, iron as building materials first appeared in Britain, but only cast iron used to be built arch bridges until 1840. With the development of the rivets connection and wrought iron technology, cast iron structure was gradually replaced by wrought iron structure, such as Brittania Bridge (in Wales, built in the 1846-1850 period)(as shown in Figure 8.4). After the inventions of the BOF (basic oxygen furnace, Britain, in 1855) and open-hearth (French, in 1865) methods, the industrialized mass production of steel is becoming possible. The steel has been the main material of the metal structure since 1890. In the early 20th century, the welding technology as well as the high-strength bolts connection (in 1934) have greatly contributed to the development of the steel structure. Gradually, steel structure has been important structural system accepted by worldwide countries.

在国外，最早将铁作为建筑材料的是英国，但是直到 1840 年才将铸铁用于拱桥。随着铆钉连接和锻铁技术的发展，铸铁结构逐渐取代了锻铁结构，如布列塔尼亚桥(威尔士，建于 1846—1850 年间)(如图 8.4 所示)。在发明转炉方法(氧气顶吹火炉，英国，1855 年)和平炉方法(法国，1865 年)之后，工业化的大规模钢铁生产已经成为可能。从 1890 年起钢成为金属结构的主要材料。20世纪早期，焊接技术以及高强度螺栓连接技术(1934)在很大程度上促进了钢结构的发展。逐渐地，钢结构已经成为重要的结构系统而被全世界的国家所接受。

Figure 8.4　Brittania Bridge in Wales
图8.4　威尔士的布列塔尼亚桥

In a long period of time from 1949 in China, the development of the steel structure in major engineering is very slow due to the constraints of steel production. With the rapid increase of national economy since 1978, steel structures have been applied wildly in our country. In recent years, a large number of steel structure fabrication and installation enterprises have emerged, and the famous foreign steel manufacturers have also broken into the Chinese market. Nowadays, according to the engineering practice and research results, Code for design of steel structures (GB50017-2003) and Technical code of cold-formed thin-wall steel structures (GB50018-2002) of China have been emended comprehensively. It is obvious that the development of steel structure will be more and more extensive in China.

中国从1949年起在很长一段时期内，由于受钢铁产量的限制，钢结构在主要工程的发展非常缓慢。自1978年以来，随着国民经济的快速增长，钢结构在我国被广泛应用。最近几年，很多著名的外国钢铁制作安装企业涌现出来，也进入了中国市场。现在，根据工程实践和研究成果，钢结构设计规范(GB50017—2003)和中国冷弯薄壁钢的技术规范(GB50018—2002)已经进行了全面修订。显而易见，中国钢结构的发展范围会越来越广泛。

## 8.2　Steel Structures Characteristic　钢结构的特点

Steel has been an important material of buildings, bridges, and other structures for more than a century. Compared with other materials, steel structure has the following characteristics.

钢成为建筑、桥梁和其他结构的重要组成部分已经超过一个世纪。与其他材料相比，钢结构具有以下特性。

(1) High strength, light weight.

强度高，重量轻。

Relative to other structural systems, steel is light weight and can reduce foundation costs. Steel has not only higher density but also higher strength than masonry and concrete material. So the weight of a steel roof is only one-third of that using concrete roof if they are under the same loading.

相对于其他结构系统，钢重量轻，可以减少基础成本。钢结构相比砌体和混凝土材料不仅密度大而且强度高,所以如果钢屋顶和混凝土屋顶在承受相同的载荷情况下,钢屋顶的重量只有混凝土的 1/3。

(2) Material uniformity, good plasticity and tenacity.

材料均匀性、良好的塑性和韧性。

Compared with masonry and concrete, the interior organizational structure of steel is uniform and isotropic approximately. Generally, the constitutive relationship of steel can be simplified as ideal elastic-plastic, which accords well with basic assumption in engineering mechanics.

与砌体和混凝土相比,钢结构的内部组织更接近匀质和各向同性体。通常地,钢的本构关系可以简化为理想弹塑性,这与工程力学中的基本假设较符合。

Typical stress-stain diagram for steel is shown in Figure 8.5.

钢的典型应力-应变图如图 8.5 所示。

Idealized stress-stain diagram for steel is shown in Figure 8.6.

钢的理想应力-应变图如图 8.6 所示。

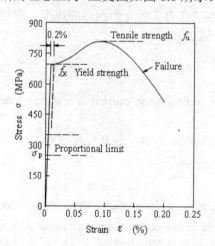
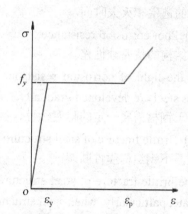

Figure 8.5　Typical Stress-stain Diagram for Steel　　Figure 8.6　Idealized Stress-stain Diagram for Steel

图 8.5　钢的典型应力-应变图　　图 8.6　钢的理想应力-应变图

(3) Easy in fabrication and good weldability.

制作简单并有良好的焊接性能。

Structural steel can be designed with large spans and bay sizes, thereby providing more flexibility in space arrangement and rearrangement.

建筑结构钢材可以用来设计大跨度结构,从而提供更多的空间并灵活布置。

(4) Good sealing.

密封良好的结构。

The steel structures with welded connections have good water tightness and air tightness. It is often applied in pressure vessels, piping, and even aerospace structures.

钢结构焊接连接有良好的水密性和气密性。经常应用于压力容器、管道,甚至航空航天结构。

(5) Reusability of this material.
材料的重复利用性。

Steel is a durable, long-lasting material and is recyclable. Steel can be easily modified and reinforced if the owner chooses to expand the facility, or if architectural changes are made.

钢是一种耐用、持久、可循环利用的材料，如果使用者选择扩建或者改变建筑设计，可以很容易地对钢材进行修改和加固。

(6) Heat resistance but non-refractory.
钢结构的耐热性好，但防火性能差。

A steel structure is easily collapsed in fire, so it is necessary to do fireproof protection on it. The most common way nowadays is fireproof painting. It can enhance the fireproofing of the steel structure and delay the rate of collapsing in fire. It also offers more time for escaping and putting out of the fire.

钢结构在火中易倒塌，所以对钢结构进行防火保护是非常必要的，目前常见的方式是在表面涂上防火涂料，它可以提高钢结构的防火性和延迟其在火中倒塌的速度。它也提供了更多的逃生和灭火时间。

(7) Poor corrosion resistance.
钢结构的抗腐蚀性差。

In the light of corrosion resistance of steel, the atmospheric corrosion resistant steel and stainless steel are developed gradually.

由于钢材需要一定的耐腐蚀性，故耐大气腐蚀钢和不锈钢也逐渐发展起来。

(8) Brittle fracture of steel structure under low temperature.
低温下钢结构脆性断裂。

The brittle fracture of steel structures is one of the most dangerous failure form under their limit state, particularly, when steel structure is subjected to low temperature and dynamic loading.

在钢结构极限状态下，脆性断裂是钢结构一个最危险的失效形式之一，在钢结构受到低温和动载作用下会特别明显。

## 8.3  Application Of Steel Structure  钢结构的应用

There are many potential benefits in the use of steel structures. For instance, steel construction can substantially reduce construction time for the frame because of off-site fabrication and the ability to construct in all seasons. This reduces on-site management and overhead costs, and improves cash flow. Careful project management and design of structural steel construction can help to ensure that above benefits achieved. In view of these advantages of steel, the application scope of the steel structures is expanding continuously. Some of these include the following.

钢结构有许多潜在的优点。例如，因为钢结构可以预制并在各季节都能建造而大大减少了建设时间。这种时间的节约减少了现场管理和间接费用，提高了资金流通性。良好的钢结构工程管理和设计有助于确保发挥上述优势。从钢材的这些优势来看，钢结构的应用规模会进一步扩大，其中包括以下几种。

1) Large span structure  大跨度结构

Because of the reasonable mechanical performance of steel and the broad developmental of the large span structure in our country, it is assured that this steel structure will be used more and more widely.

在我国，因为钢具有合理的机械性能和大跨度结构的广泛发展空间，可以肯定的是钢结构的使用范围会越来越广泛。

National swimming centre "water cube" is shown in Figure 8.7.

如图 8.7 扭不为国家游泳中心水立方。

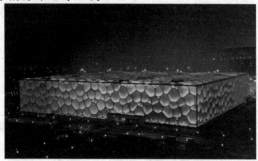

Figure 8.7  National Swimming Center "Water Cube"

图 8.7  国家游泳中心水立方

Roof of Harbin meeting and exhibition center is shown in Figure 8.8.

如图 8.8 所示为哈尔滨会议展览中心屋顶。

Figure 8.8  Roof of Harbin Meeting and Exhibition Center

图 8.8  哈尔滨会议展览中心屋顶

2) Industrial building  工业建筑

The steel structure is usually adopted in the framework of construction with large weight crane. In recent years, due to the profiled steel sheets in application to roof system engineering, industrial building has been under rapid development.

钢结构通常应用于带有大重量吊车的排架工业建筑中。近年来，由于异形钢板在屋面系统工程的应用，工业建筑得到快速发展。

Gable portal frame structure is shown in Figure 8.9.

如图 8.9 所示为山墙门户框架结构。

Figure 8.9　Gable Portal Frame Structure

图 8.9　山墙门户框架结构

3) Building under repeated loads and dynamic loads　在循环荷载和动荷载下的建筑

Due to the good plasticity and tenacity of steel, the steel structure is more appropriate for the structure under dynamic loads.

良好的塑性和韧性使得钢结构更适合应用于动荷载作用下的结构。

4) High-rise and multi-story building　高层和多层建筑

Actually, the early development of high-rise buildings began with structural steel frame. The main forms of steel high-rise and multi-story building are, frame structure, braced-frames structure, frame-tube structure, suspended structure, and mega-frame structure etc, as shown in Figure 8.10 and Figure 8.11.

事实上，高层建筑的早期发展是以钢结构框架开始的。钢高层和多层建筑的主要形式是框架结构、支撑框架结构、组合框架结构、悬挂结构、组合结构等，如图 8.10 和图 8.11 所示。

Figure 8.10　Jin Mao Building of Shanghai and Shanghai world Financial Center

图 8.10　上海金茂大厦和上海环球金融中心

Figure 8.11　Bank of China Building of Hong Kong

图 8.11　香港中国银行

5) Tower structure　塔结构

The tower structure includes the tower truss and guyed mast，such as tower trusses for high voltage transmission line or broadcast, the television tower, the mast for communication, and the tall rocket launch tower, as shown in Figure 8.12 and Figure 8.13.

塔架结构包括塔桁架和拉线式桅杆，如高压输电线路、广播电视塔、信号塔和高耸的火箭发射塔，如图 8.12 和图 8.13 所示。

Figure 8.12　TV Tower of Guang zhou

图 8.12　广州电视塔

Figure 8.13　The Rocket Launch Tower

图 8.13　火箭发射塔

6) Demountable structure　可拆卸结构

Demountable structure generally includes steel scaffolding (as shown in Figure 8.14), bamboo wooden scaffolding scaffolding etc.

可拆卸结构一般包括钢脚手架(如图 8.14 所示)，木脚手架、竹脚手架等。

7) Vessel and else structure　容器及其余结构

This refers to the pressure vessel, the oil tank, the hot blast stove and offshore production platform (as shown in Figure 8.15) etc.

这些指的是压力容器、油舱、热风炉和海上生产平台(如图 8.15 所示)等。

Figure 8.14　The steel scaffold

图 8.14　钢脚手架

Figure 8.15　The offshore production platform

图 8.15　海上生产平台

8) Light steel structure　轻钢结构

The type of light steel structure includes portal frame with tapered members, cold-formed thin-wall steel structure and steel tube structure, as shown in Figure 8.16 and Figure 8.17.

轻钢结构的类型包括变截面构件的门式框架，冷弯薄壁钢结构和钢管结构，如图 8.16 和图 8.17 所示。

Figure 8.16　The Welding Tube Structure (Nan jing)　　Figure 8.17　The Cold-Formed Thin-Wall Steel Structure
图 8.16　焊接管结构(南京)　　　　　　　　　　图 8.17　冷弯薄壁钢结构

9) Steel-concrete composite structure　钢-混凝土组合结构

In the recent years, the systems combining both concrete and steel have also been developed, which combines the advantages of both reinforced concrete and structural steel systems. The Saiger Plaza in Shenzhen is based on this system, as shown in Figure 8.18.

近年来，混凝土和钢的组合系统也已经被开发出来，它结合了钢筋混凝土和钢结构系统两者的优点，深圳赛格广场就是这样的结构系统，如图 8.18 所示。

Figure 8.18　Saiger Plaza of Shenzhen
图 8.18　深圳赛格广场

## 8.4　Aim of Steel Structural Design　钢结构的设计目标

The aim of structural design should be to provide, with due regard to economy, a structure capable of fulfilling its intended function and sustaining the specified loads for its intended life. The design should facilitate safe fabrication, transport, handling and erection. It should also take account of the needs of future maintenance, final demolition, recycling and reuse of materials.

钢结构设计的目标除了应该适当考虑其经济性，还有要实现其设计功能的结构承载力和支撑其在设计生命周期内的某些荷载。设计应该方便安全地制造、运输、安装和建造。它也应该考虑未来的维护、使用结束后的拆除、材料的回收和再利用等需求。

The structure should be designed to behave as a one three-dimensional entity. The layout of its constituent parts, such as foundations, steelwork, joints and other structural components should constitute a robust and stable structure under normal loading to ensure that, in the event of misuse or accident, damage will not be disproportionate to the cause.

结构应按照一个三维实体去设计。结构的组成部分，如基础、钢结构构件、节点等应设计成一个在正常荷载下坚固而稳定的结构，以确保在错误使用或发生意外事件的情况下，不致因此造成不成比例的破坏。

To achieve these aims the basic anatomy of the structure by which the loads are transmitted to the foundations should be clearly defined. Any features of the structure that have a critical influence on its overall stability should be identified and taken account of in the design.

为了实现这些目标，对结构的荷载传递到基础的基本分析应该清楚地加以确定。结构的任何构件对整体的稳定性都有很关键的影响，这应该得到重视并在设计中予以考虑。

## 8.5　Limit State　极限状态

For most of calculations in steel structures except fatigue calculation, the limit state design method based on probabilistic theory is adopted, using design expressions with partial safety factors. For the problem of fatigue calculation, the allowable stress design method is used.

除了疲劳计算外，对于钢结构的大多数计算一般采用基于概率理论的极限状态设计方法，即采用部分安全系数设计表达式来计算。对于疲劳计算的问题，采用容许应力法设计。

Limit state design considers the functional limits of strength, stability and serviceability of both single structural elements and the structure as a whole. Design shall be based on the principle that no applicable strength or serviceability limit state shall be exceeded when the structure is subjected to all appropriate load combinations. Load-carrying structures shall be designed according to the following ultimate limit states and serviceability limit states.

极限状态设计考虑单一的结构构件和结构整体的强度、稳定性和使用性的功能极限。设计时应依据的原则是：当结构受到所有荷载组合作用时，结构强度或使用性不会超过极限状态。承载结构的设计应根据承载能力极限状态和正常使用极限状态进行。

Ultimate limit states consider the safety of the whole or part of the structure. Examples of ultimate limit states are strength including yielding, rupture, buckling and forming a mechanism against overturning, sliding, uplift and overall lateral or torsional sway buckling, fire leading to deterioration of mechanical properties at elevated temperatures and thermal actions, fracture caused by brittle material behaviour or by fatigue.

承载能力极限状态考虑全部或部分结构的安全性。它的例子有：包括屈服、破裂、强化和形成机构时的强度，包括抗倾覆、滑动、上升和整体侧移或扭转变形的稳定，造成结构在高温下力学性能破坏的火灾，由于材料脆性和疲劳导致的破坏。

1) The ultimate limit states include the following:
极限状态包括以下几种：

(1) Strength failure of members and connections.
构件或者连接强度失效。

(2) Fatigue failure.
疲劳失效。

(3) Excessive deformation no longer suitable for carrying load.
变形过大不适宜继续承载。

(4) Loss of stability of structures and members.
结构或者构件的失稳。

(5) Formation of mechanism.
塑性铰的形成。

(6) Overturning of the structure.
结构的倾覆。

Serviceability limit states correspond to limits beyond which specified in-service criteria are no longer met. Examples are deflection, wind induced oscillation, human induced vibration and durability.
承载力极限状态指的是超过某个标准不再满足条件的极限。如变形、风致振动、人类引起的振动和耐久性。

2) The serviceability limit states include the following:
正常使用极限状态包括以下几种：

(1) Deformations affecting normal use and appearance of a structure, structural and non-structural components.
影响结构或结构构件、非结构构件正常使用或外观的变形。

(2) Vibration affecting normal use.
影响正常使用的振动。

(3) Local damage (including concrete cracks) affecting normal use or durability.
影响正常使用或耐久性的局部损坏(包括混凝土裂缝)。

## 8.6 Load and Calculation of Load Effects
荷载与荷载效应计算

All relevant loads should be considered separately and in such realistic combinations as to comprise the most critical effects on the elements and the structure as a whole. The magnitude and frequency of fluctuating loads should also be considered. Loading conditions during erection should receive particular attention. Settlement of supports should be taken into account where necessary.
所有相关的荷载都应该单独考虑，并考虑实际情况做组合。这样做是为了包括对构件和结构整体构成的最大影响。变化的荷载幅度和频率都应该考虑。加载过程的情况应受到特别的重视。在必要的情况下应考虑支座沉降的问题。

In the Chinese Code for design of steel structures, there are such provisions as follows:
在中国钢结构设计规范中有如下规定：

(1) In designing a steel structure according to the ultimate limit state, the basic combination of load effects shall be considered and, if necessary, the accidental combination of load effects shall also be considered.

根据极限状态设计钢结构时，应当考虑荷载效应的基本组合，如果有必要，也应当考虑偶然荷载效应的组合。

(2) In designing a steel structure according to the serviceability limit state, the normal combination of load effects shall be considered, whereas for composite steel and concrete beams, the quasi-permanent combination shall also be considered.

当根据正常使用极限状态设计钢结构时，应考虑荷载的标准组合，而对复合钢和混凝土梁，应该考虑荷载准永久组合。

(3) In checking the strength and stability of structures or structural members and also the strength of connections, the design value of loads shall be used (i.e. the characteristic value of loads multiplied by partial safety factor for loads), whereas in checking fatigue, the characteristic value of loads shall be used.

在检查结构或结构构件的强度、稳定性和连接强度时，应当使用荷载的设计值(即荷载分项安全系数乘以荷载的特征值)；而在检查疲劳时，应使用荷载的特征值。

(4) In the design of steel structures, the characteristic value of loads, the partial safety factor for loads, the load combination coefficient, the impact factor of dynamic loads shall comply with the requirements of the current national standard.

在设计钢结构时，荷载的特征值、荷载分项安全系数、荷载组合系数、动态载荷的影响系数应符合现行国家标准的要求。

## 8.7　Material Selection　材料选择

The grade and quality of steel used for load-carrying structures shall be selected in taking comprehensive account of the importance, the loading characteristic, the structural type, the stress state, the connection device, the thickness of steel and the working circumstance of the structure, etc. The material of load-carrying structures may be steel of grades Q235, Q345, Q390 and Q420.

选择用于承载结构的钢的等级和质量时应综合考虑结构的重要性、荷载特性、结构类型、应力状态、连接装置、钢片的厚度和结构的工作情况等。承载结构的材料可以是等级为 Q235、Q345、Q390 和 Q420 的钢。

Steel for load-carrying structures shall be guaranteed for meeting the requirements of tensile strength, percentage of elongation, yield strength (or yield point) and also of proper sulfur and phosphorus contents. For welded structures, the steel shall also be guaranteed for proper carbon content. The steel for welded and important non-welded load-carrying structures shall also be guaranteed for passing the cold-bending test.

钢承载结构应保证满足抗拉强度、伸长率、屈服强度(屈服点)以及适当的硫和磷含量的要求。对于焊接结构，应保证钢中适当的碳含量。用于焊接和重要的非焊接承载结构的钢，也应保证通过冷弯曲试验。

## 8.8 Design Indices 强度设计值

The design value of steel strength shall be taken from Table 8.1 according to the steel thickness or diameter. For the design value of strength of cast steel parts and connection strength, there are also corresponding provisions.

钢的强度设计值，应采用如表 8.1 所示的钢材厚度或直径。铸钢件的强度和连接强度设计值也有相应的规定。

Table 8.1 Design Value of Steel Strength (N/mm$^2$)

表 8.1 钢材强度设计值(N/mm$^2$)

| Steel 钢 | | Tension, compression and bending $f$ 抗拉、抗压和抗弯 $f$ | Shear $f_v$ 抗剪 $f_v$ | End bearing (planed and closely fitted) $f_{ce}$ 端轴承(位置固定且连接紧密) $f_{ce}$ |
|---|---|---|---|---|
| Grade 等级 | Thickness or diameter 厚度或直径(mm) | | | |
| Q235 | ≤16 | 215 | 125 | 325 |
| | >16~40 | 205 | 120 | |
| | >40~60 | 200 | 115 | |
| | >60~100 | 190 | 110 | |
| Q345 | ≤16 | 310 | 180 | 400 |
| | >16~35 | 295 | 170 | |
| | >35~50 | 265 | 155 | |
| | >50~100 | 250 | 145 | |
| Q390 | ≤16 | 350 | 205 | 415 |
| | >16~35 | 335 | 190 | |
| | >35~50 | 315 | 180 | |
| | >50~100 | 295 | 170 | |
| Q420 | ≤16 | 380 | 220 | 440 |
| | >16~35 | 360 | 210 | |
| | >35~50 | 340 | 195 | |
| | >50~100 | 325 | 185 | |

Note: Thickness in this table denotes the steel thickness at the calculation location, for members subject to axial force, it is the thickness of the thicker plate element of the section.

注：表中厚度是指计算点的厚度，对轴心受力构件是指截面中较厚板件的厚度。

## 8.9 Provisions for Deformation of Structures and Structural Members 结构及构件变形限值

The deflections of a building or part under serviceability loads should not impair the strength or efficiency of the structure or its components, nor cause damage to the finishing. In order not to impair the serviceability, nor to affect the appearance of structures and structural members, their deformation (deflection or lateral drift) shall comply with the relevant limiting values in designing. When checking for deflections the most adverse realistic combination and arrangement of serviceability loads should be assumed, and the structure may be assumed to behave elastically.

在正常荷载下的建筑整体或部分的挠度应不影响结构或组成部件的强度或效率，也不会对最后的结构造成破坏。为了不影响适用性，同时不影响结构和结构构件的外观，其变形(偏移或横向偏移)应符合设计相关的限值。当检查挠度时，应考虑结构在最不利实际荷载的组合和布置的条件，该结构可以被假定为具有弹性。

The allowable values of deformation, as a general rule, are specified in Appendix of Chinese Code. The values therein may be suitably modified in consideration of practical experiences or to meet a specific demand, provided the serviceability is not impaired nor the appearance affected.

作为一般规则，允许变形值在中国规范的附录中记录。只要其使用性不受影响或外观不受影响，其中的值可以根据实践经验或为了满足特定的需求而做出适当的修改。

The deflection of crane girders, floor beams, roof girders, working platform beams and members of wall framing should not exceed the relevant allowable value listed in Table 8.2.

吊车梁、梁、屋面梁、工作平台梁和墙框架的挠度应不超过如表 8.2 中所列出的有关允许值。

Table 8.2  Allowable Deflection of Flexural Members

表 8.2  受弯构件的挠度容许值

| Item No. 项次 | Type of member 构件类别 | Allowable deflection 挠度容许值 | |
|---|---|---|---|
| | | $[v_T]$ | $[v_Q]$ |
| 1 | Crane girders and crane trusses (calculated deflection under self-weight and one of the cranes of largest capacity) 吊车梁和吊车桁架(按自重和起重量最大的一台吊车计算挠度) | | |
| | (1) for hand operated cranes and monorails (including under slung cranes) 手动吊车和单梁吊车(含悬挂吊车) | $l/500$ | — |
| | (2) for bridge cranes of light duty 轻级工作制桥式吊车 | $l/800$ | |
| | (3) for bridge cranes of medium duty 中级工作制桥式吊车 | $l/1000$ | |
| | (4) for bridge cranes of heavy duty 重级工作制桥式吊车 | $l/1200$ | |

Continued 续表

| Item No. 项次 | Type of member 构件类别 | Allowable deflection 挠度容许值 | |
|---|---|---|---|
| | | $[v_T]$ | $[v_Q]$ |
| 2 | Beam-rails for hand operated or electric hoists 手动或电动葫芦的轨道梁 | $l/400$ | — |
| 3 | Beams of working platform under track with heavy rail(weighing 38kg/m or more) 有重轨(重量等于或大于38kg/m)轨道的工作平台梁 | $l/600$ | — |
| | Beams of working platform under track with light rail(weighing 24kg/m or less) 有轻轨(重量等于或小于24kg/m)轨道的工作平台梁 | $l/400$ | — |
| 4 | Floor(roof) beams or trusses, platform beams(except Item No.3) and platform slabs 楼(屋)盖梁或桁架，工作平台梁(第3项除外)和平台板 | | |
| | (1) main girders or trusses (including those with under slung hoisting equipment) 主梁或桁架(包括设有悬挂起重设备的梁和桁架) | $l/400$ | $l/500$ |
| | (2) beams with plastered ceiling 抹灰顶棚的次梁 | $l/250$ | $l/350$ |
| | (3) beams other than Item No.(1) and (2) (including stair beams) 除(1)、(2)款外的其他梁(包括楼梯梁) | $l/150$ | $l/300$ |
| | (4) purlins under roofing of corrugated iron and asbestos sheet with no dust accumulation profiled metal sheet, corrugated iron and asbestos sheet etc. with dust accumulation other material 屋盖檩条支撑无积灰的瓦楞铁和石棉瓦屋面的； 支撑压型金属板、有积灰的瓦楞铁和石棉瓦等屋面以及支撑其他屋面材料的 | $l/200$ | — |
| | (5) platform slabs 平台板 | $l/150$ | — |
| 5 | Members of wall framing (taking no account of gust coefficient for wind load) 墙架构件（风荷载不考虑阵风系数） | | |
| | (1) stud 支柱 | — | $l/400$ |
| | (2) wind truss (acting as support to continuous stud) 抗风桁架(作为连续支柱的支撑时) | — | $l/1000$ |
| | (3) girt (horizontal) in masonry wall 砌体墙的横梁(水平方向) | — | $l/300$ |
| | (4) girt (horizontal) for cladding of profiled metal sheet, corrugated iron and asbestos sheet 支承压型金属板、瓦楞铁和石棉瓦墙面的横梁(水平方向) | — | $l/200$ |
| | (5) girt (vertical and horizontal) for glass window 带有玻璃窗的横梁(竖直和水平方向) | $l/200$ | $l/200$ |

Note: ① $l$ denotes the span length of a flexural member (for cantilever beam and overhanging beam, it is twofold the overhang).
② $[v_T]$ is the allowable deflection under the total unfactored load (the camber shall be deducted if there exists any).
③ $[v_Q]$ is the allowable deflection under the unfactored variable load.

注：① $l$ 为受弯构件的跨度(对悬臂梁和伸臂梁为悬伸长度的2倍)。
② $[v_T]$ 为永久荷载和可变荷载标准值产生的挠度(如有起拱应减去挠度)的容许值。
③ $[v_Q]$ 为可变荷载标准值产生的挠度的容许值。

The horizontal displacement of the column top and the story drift in a frame should not exceed the following values under the action of unfactored wind load.

在风荷载的作用下梁顶端的水平位移和楼层的位移不应该超过以下值：

(1) Column top displacement of single story frame without bridge crane $H/150$.

没有装配桥式起重机的单层的框架结构，柱顶的位移不应超过其高度的一百五十分之一。

(2) Column top displacement of single story frame equipped with bridge crane $H/400$.

装配有桥式起重机的单层的框架结构柱顶的位移不应超过其高度的四百分之一。

(3) Column top displacement of multistory frame $H/500$.

多层的框架结构，柱顶的位移不应超过其高度的五百分之一。

(4) Story drift of multistory frame $h/400$.

多层框架楼层位移不应超过其高度的四百分之一。

$H$ is the total height from top of foundation to column top; $h$ is the story height.

$H$ 代表从基础到柱顶的整个高度，$h$ 代表楼层高。

Note: ① For multistory framed structure of a civil building requiring refined indoor decoration, the story drift should be reduced appropriately. For multistory frames without wall, the story drift may be appropriately enlarged.

② For light frame structures, both the column top displacement and the story drift may be enlarged appropriately.

注：① 对民用的多层框架结构而言，室内的装饰需要重新定义，楼层的允许位移也需要适当地减少，对于没有隔墙的多层框架来说，楼层位移可以适当放宽要求。

② 对于轻型框架结构来说，无论是柱顶位移还是层间位移都可以适当地放宽要求。

## New Words and Expressions 生词和短语

1. Steel Structure 钢结构
2. beam  *n.* 横梁
3. girder  *n.* 纵梁
4. bar  *n.* 条，棒
5. plate  *n.* 平板
6. span  *n.* 跨度，跨距；范围
7. connection  *n.* 连接；关系；连接件
8. uniformity  *n.* 均匀性；一致；同样
9. plasticity  *n.* 塑性，可塑性；适应性；柔软性
10. tenacity  *n.* 韧性；固执；不屈不挠；黏性
11. isotropic  *adj.* 各向同性的；等方性的
12. weldability  *n.* 焊接性；可焊性
13. flexibility  *n.* 灵活性；弹性；适应性
14. good sealing  良好密封性

15. reusability    n. 可重用性
16. heat resistance    耐热性
17. corrosion resistance    耐蚀性；抗腐蚀性
18. brittle fracture    脆性破坏
19. repeated load    重复载荷
20. dynamic load    动力载荷
21. guyed mast    拉线式电杆
22. voltage    n. [电] 电压
23. demountable structure    可移动结构
24. erection    n. 建造；建筑物；直立
25. layout    n. 布局；设计；安排；陈列
26. serviceability    n. 可用性，适用性；使用可靠性
27. load-carrying structure    承力结构
28. sliding    n. 滑；移动
              adj. 变化的；滑行的
29. torsional    adj. 扭转的；扭力的
30. criteria    n. 标准，条件
31. oscillation    n. 振荡；振动；摆动
32. durability    n. 耐久性；坚固；耐用年限
33. provision    n. 规定；条款；准备
34. quasi-permanent combination    准永久组合
35. fatigue    n. 疲劳，疲乏
36. load combination coefficient    荷载组合系数
37. percentage of elongation    延伸率
38. yield strength    屈服强度
39. computation    n. 估计，计算
40. nonlinear    adj. 非线性的
41. inelastic    adj. 无弹性的；无适应性的；不能适应的
42. stability    n. 稳定性；坚定，恒心
43. stiffness    n. 僵硬；坚硬；不自然；顽固
44. deformation    n. 变形
45. statically    adv. 静态地；静止地
46. equivalent    n. 等价物，相等物
47. inelasticity    n. 不适应性；坚硬性；无弹力
48. residual stress    剩余应力

# Exercises 练习

I. Write a T in front of a statement if it is true according to the text and write an F if it is false.

1. Relative to other structural systems, steel is light weight and can reduce foundation costs.

2. In the light of had corrosion resistance of steel, the atmospheric corrosion resistant steel and stainless steel are developed gradually.

3. For the problem of fatigue calculation, the allowable stress design method is used.

4. Limit state design shall be based on the principle that all the applicable strength or serviceability limit state should be exceeded when the structure is subjected to all appropriate load combinations.

5. In designing a steel structure according to the ultimate limit state, the basic combination of load effects shall not be considered.

6. The deflections of a building or part under serviceability loads should not impair the strength or efficiency of the structure or its components, nor cause damage to the finishing.

II. Complete the following sentences.

1. Please list five steel structure characteristics: _____.

2. The tower structure includes _____.

3. In designing a steel structure according to the serviceability limit state, _____ shall be considered, whereas for composite steel and concrete beams, _____ shall also be considered.

4. The material of load-carrying structures may be steel of grades _____.

5. For light frame structures, both the column top displacement and the story drift may be _____ appropriately.

III. Translate the following sentences into Chinese.

1. Steel is an alloy of iron-carbon, the history of steel structure application is closely related to the development of iron-making and steel-making technology.

2. Generally, the constitutive relationship of steel can be simplified as ideal elastic-plastic, which accords well with basic assumption in engineering mechanics.

3. The brittle fracture of steel structures is one of the most dangerous failure forms under their limit state, particularly, when steel structure is subjected to low temperature and dynamic loading.

4. Because of the reasonable mechanical performance of steel structure and the broad developmental space of the large span structure in our country, it is sure that this steel structure will be used more and more widely.

5. For the most of calculations in steel structures except fatigue calculation, the limit state

design method based on probabilistic theory is adopted, using design expressions with partial safety factors.

6. In designing a steel structure according to the ultimate limit state, the basic combination of load effects shall be considered and, if necessary, the accidental combination of load effects shall also be considered.

Ⅳ. Translate the following sentences into English.

1. 在国外，铁作为建筑材料首先出现在英国，但直到 1840 年铸铁才用于建造拱桥。
2. 中国从 1949 年开始在很长的一段时间里，由于钢材生产的限制，钢结构在重大工程中的发展非常缓慢。
3. 钢结构在火灾中容易倒塌，因此需要对它进行防火保护。
4. 鉴于钢结构的这些优点，钢结构的应用范围在不断地扩大。
5. 由于钢的良好的塑性和韧性，对于动态荷载作用下的结构，钢结构更为合适。
6. 最大极限状态考虑全部或部分结构的安全。

# Answers 答案

Ⅰ.

1. T  2. T  3. T  4. F  5. F  6. T

Ⅱ.

1. High strength, light weight; Material uniformity, good plasticity and tenacity; Easy in fabrication and good weldability; Good sealing; Reusability

2. tower truss; guyed mast

3. the normal combination of load effects; the quasi-permanent combination

4. Q235, Q345, Q390 and Q420

5. enlarged

Ⅲ.

1. 钢是铁碳合金，钢结构应用的历史与炼铁和炼钢技术的发展密切相关。
2. 一般来说，钢的本构关系可以简化为理想弹塑性，这与工程力学的基本假设相吻合。
3. 钢结构的脆性断裂是钢结构在极限状态下最危险的失效形式之一，特别是当钢结构受到低温和动态加载时。
4. 由于钢结构的合理的机械性能和在我国大跨度结构的广阔的发展空间，很明显钢结构将被越来越广泛地使用。
5. 除了疲劳计算之外，对于大多数钢结构的计算，基于概率理论基础上的极限状态设计方法利用局部安全系数设计表达式。
6. 根据最大极限状态法设计钢结构时，基本荷载组合作用应该被考虑，如果有必要，偶然荷载组合作用也应予以考虑。

## IV.

1. In foreign countries, iron as building materials first appeared in Britain, but only cast iron used to be built arch bridges until 1840.

2. In a long period of time from 1949 in China, the development of the steel structure in major engineering is very slow due to the constraints of steel production.

3. A steel structure is easily collapsed in fire, so it is necessary to do fireproof protection on it.

4. In view of these advantages of steel structures, the application scope of the steel structures is expanding continuously.

5. Due to the good plasticity and tenacity of steel, the steel structure is more appropriate for the structure under dynamic loads.

6. Ultimate limit states consider the safety of the whole or part of the structure.

# Chapter 9　Masonry Structure
# 第 9 章　砌体结构

Masonry construction has been used for at least 10,000 years in a variety of structures—homes, private and public buildings and historical monuments. The masonry of ancient times involved two major materials: brick manufactured from sun-dried mud or burned clay and shale; and natural stone.

砌体结构在各种房屋结构中至少已使用了 1 万年，如私人、公共建筑及纪念碑等。古代的砌体主要涉及两种材料：由晒干的泥或烧黏土和页岩制成的砖、天然石材。

The first masonry structures were unreinforced and intended to support mainly gravity loads. The weight of these structures stabilized them against lateral loads from wind and earthquakes.

最先的砌体结构是无钢筋的，而且主要承受重力荷载，依靠自身的重量来对抗来自风和地震的水平荷载。

Masonry construction has progressed through several stages of development. Fired clay brick became the principal building material in the United States during the middle 1800s. Concrete masonry was introduced to construction during the early 1900s and, along with clay masonry, expanded in use to all types of structures. Historically, "rules of thumb" (now termed "empirical design") were the only available methods of masonry design. Only in recent times have masonry structures been engineered using structural calculations. In the last 45 years, the introduction of engineered reinforced masonry has resulted in structures that are stronger and more stable against lateral loads, such as wind and seismic.

砌体结构的发展经历了若干个阶段。19 世纪中期，烧制黏土砖成为美国主要的建筑材料。20 世纪初，混凝土砌块被引入建筑中，而且和黏土砌块一样扩大应用于所有类型的结构上。从历史上看，"拇指规则"(现称为"经验设计")是砌体设计唯一可用的方法。只有在近代才对砌体结构进行结构计算，并在此基础上进行设计，在过去的45 年中，设计并使用配筋砌体的结果是结构能抵抗更强的水平荷载(如风和地震)并更加稳定。

Masonry consists of a variety of materials. Raw materials are made into masonry units of different sizes and shapes, each having specific physical and mechanical properties. Both the raw materials and the method of manufacture affect masonry unit properties. The word "masonry" is a general term that applies to construction using hand-placed units of clay, concrete, structural clay tile, glass block, natural stones and the like. One or more types of masonry units are bonded together with mortar, metal ties, reinforcement and accessories to form walls and other structural elements. Proper masonry construction depends on correct design, materials, handling, installation and workmanship. With a fundamental understanding of the functions and properties of the materials that comprise masonry construction and with proper design and construction, quality masonry structures are not difficult to obtain.

砌体包括多种材料。原材料被制成不同尺寸和形状的砌体构件，每个构件都具有特定的物理和机械性能。原材料和制造方法都会影响砌体的单元属性。"砌体"这个词是一个通用的术语，适用于建筑用手工放置单元的黏土砖、混凝土、结构黏土瓦、玻璃块、天然石等。一种或多种类型的砌块由砂浆、配筋、加固配件等结合在一起形成墙体和其他构件。合理的砌体结构取决于正确的设计、专用的材料、运输以及正确的安装和工艺。对组成砌体结构的材料性能与功能有了一个基本的了解后，再加上适宜的设计和施工，高质量的砌体结构是不难得到的。

## 9.1　Introduction　引言

Clay brick masonry building is the most likely used type of structural system on housing in Peru and South America. More than 43% of housing are built using this system. In last 2001 Atico (Southern Peru) quake housing build with masonry damage. Main reason of this damage is the non quality control on the construction and improper structural configuration. Building a house without following the National Standards of Earthquake design, the Masonry design standard and this Masonry Construction guide could produce damage on the house.

黏土砖的砌体房屋在秘鲁和南美是最有可能使用的结构体系类型，43%以上的住房都采用这种体系建造。2001年阿蒂科(秘鲁南部)地震导致建设住房出现砖石的损伤，这种损伤的主要原因是施工不当和结构形式的不良设计。不按照抗震设计国家标准、砌体设计标准和砌体结构指南去建造房子，将来可能会对房子产生损坏。

## 9.2　Materials to Use　材料的使用

1. Cement(as shown in Figure 9.1(a))　水泥(如图9.1(a)所示)

It must be protected from humidity for not harden before its use. Storage space should be insulated from soil humidity through plastic sheets or wood stands.

水泥在使用前必须采取保护措施，使之不受潮变硬，且在有效期内使用。存储地点应该使用塑料布或木桩与地面绝缘，使之不受潮。

2. Sand (fine aggregates: fine and thick sand (as shown in Figure 9.1(b)))　砂(细骨料：细砂和粗砂如图9.1(b)所示)

It will use on the mix with cement, stone and water. Its mission is to reduce voids between stones. Sand shouldn't contain earth (soil), mica, salt, organic filthy, odor, iron compounds, blackish appearance. You can prove if sand is bad putting sand in a recipient with water. If too much soil or dust is present, it will separate from the mix.

砂与水泥、石头和水混合使用，其作用是减少石头之间的空隙。砂中不能含有土(泥)、云母、盐、有机杂质、气体、铁化合物，且表面不能变黑。如何证明砂的质量？可以把砂放入带水的容器中，如果有太多的泥土或灰尘存在，它们将从混合物中分离。

(a) Cement 水泥

(b) Sand 砂

Figure 9.1 Cement and Sand

图 9.1 水泥和砂

### 3. Crushed Stone (thick aggregates) 碎石(粗骨料)

Stone should be crushed or angular (sharp) as shown in Figure 9.2(a). Should be hard and compact. Stones easily breakable are not good.

石头应该被碾碎或是小块(小石)的(如图 9.2(a)所示)，应该坚硬和紧凑。很容易碎的石头是不好的。

### 4. Sand and Gravel (natural mix of aggregates) 砂砾石(自然混合骨料)

Sand and gravel (as shown in Figure 9.2(b)) is a natural mix of stone of different sizes and thick sand. It is used to prepare concrete of low resistance or quality like over-footing.

砂砾石(如图 9.2(b)所示)用大小不同的石头和厚砂自然混合而成，它是用来配制低流动性的混凝土，或者做基础使用。

(a) Crushed Stone 碎石

(b) Sand and Gravel (Natural Mix of Aggregates)
砂砾石(自然混合骨料)

Figure 9.2 Crush Stone, Sand and Gravel

图 9.2 碎石和砂砾石

## 5. Water  水

Water shouldn't contain filthy elements, should be clean, drinkable and fresh.
水应该是干净的、可以饮用的和新鲜的，不能含有肮脏的成分。

## 6. Masonry Units (as shown in Figure 9.3)  砌块(如图9.3所示)

Masonry units can be solid, hollow or tubular. Solid section without holes must be more than 75% of the geometrical area. The minimum compression stress of bricks is 50 kgf/cm$^2$.

砌块可以是黏土砖和石灰硅质砖。它们可以是实心的、空心的或管状的。砌块无孔部分必须是几何面积的75%以上，砖的最小压缩强度是50kgf/cm$^2$。

Figure 9.3  Concrete; Masonry Units

图9.3  混凝土和砌块

## 7. Steel Reinforcement (as shown in Figure 9.4(a))  钢筋(如图9.4(a)所示)

For confined reinforced concrete elements corrugated bars of 9.15m length and diameters of 3/8″ and 1/2″ should be used. For stirrups or hoops can be used flat bars of 1/4″ diameter. To prevent oxidation storage of bars can be cover by plastic sheets or wood boards.

对于受约束的钢筋混凝土构件，应使用长9.15米、直径为3/8英寸和1/2英寸的肋形钢筋。对于箍筋或箍可用1/4英寸直径的条形钢筋。为了防止氧化，储存的钢筋可以盖塑料布或木板。

## 8. Wood(as shown in Figure 9.4(b))  木材(如图9.4(b)所示)

Wood boards and braces are used as form (mold). Forms should be dry and protected from water; otherwise it remains humid (wet), swells up and becomes soft. It is used to apply a cover of oil (petroleum) in the surface of wood board before its use as a form.

木材板和支撑都可用作模板(模具)。模板应该是干燥和防水的；否则在潮湿时会变得膨胀和柔软。模板在使用前，常在表面涂抹一层油(石油)。

(a) Steel Reinforcement　钢筋　　　　　　　(b) Wood　木材

Figure 9.4　Steel Reinforcement and Wood

图 9.4　钢筋和木材

## 9.3　Elements are Part of The Structural System
## 　　　结构系统的组成

Elements are part of the structural system, as shown in Figure 9.5.

结构系统的组成，如图 9.5 所示。

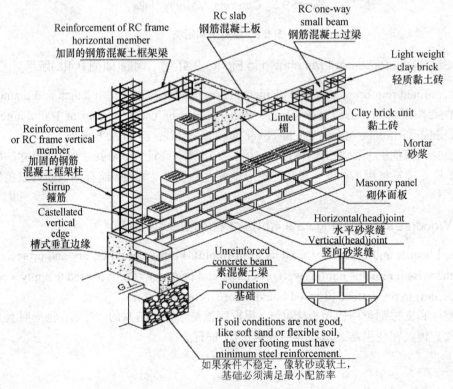

Figure 9.5　Elements Are Part of the Structural System

图 9.5　结构系统的组成

## 9.4　Preparation before Starting the Construction
## 　　　施工前准备

### 1. Preparing the Ground　场地准备

Ground should be clean, without rubbish, neither organic material nor any odd element to the ground.

场地应干净，无垃圾，地面上没有有机材料和杂物。

### 2. Drawing the structure on the ground　在场地内绘制结构轮廓

Ropes (cord) are tightened, using trestles made by wood poles nailed to a transversal stick and embedded to the ground, as shown in Figure 9.6. Trestles are placed at external part of build. Check the angle of 90 at the corners making triangle of 3-4-5 length sides, as shown here.

如图 9.6 所示，将绳收紧，用木杆制成的支架钉在横棍上。然后嵌入地面。支架放置在外部建设。在角落利用勾股定理的 3-4-5 长度的边来校验 90 度角。

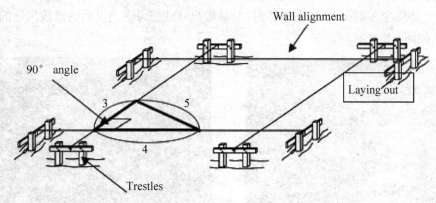

Figure 9.6　Drawing the Structure on the Ground
图 9.6　在场地内绘制结构轮廓

## 9.5　How to Build the Foundation　基础施工

### 1. Site Conditions　场地条件

The behavior of any foundation depends on the ground condition. Dense gravel, compact sand or silt or rigid clay are example of good ground. Foundations settled on these soils are expected not to experience any problem.

基础的建造都要依靠相应的地质条件。密集的砾石、紧密的砂土或者粉土、坚硬的黏土都是较好的地质条件。建在这些土壤上的基础一般不会出现任何问题。

However, if non-controlled landfill or garbage deposits compose the ground, construction on these kinds of soils is prohibited.

但若土体是未经过处理的垃圾形成，这类土体是不允许作为地基的。

### 2. Digging of Trench 沟槽开挖

A trench digging for continuous foundation should be made following the structural plans and details, as shown in Figure 9.7.

连续基础的沟槽开挖应按照图 9.7 所示的细节进行。

It is important that foundation to be leveled below the ground level, on natural soil at a depth not less than 1.0 m. If thickness of the shallow landfill is greater than 1.0 m the trench should be over excavated until it reach the natural soil and refilled with simple concrete.

基础应该在地表以下和自然土上面，埋深不小于 1m。如果土体中，浅埋的垃圾深度超过 1m，沟槽应该开挖到自然土，然后用混凝土回填。

### 3. Prepare Bottom of Foundation 准备基础底面

Bottom of trench should be compacted and leveled. Foundation dimensions should consider future expansion of the building like the increasing of the stories, at the time of the design, as shown in Figure 9.8.

基槽的垫层应压实和水平。在设计时，对基础尺寸就应考虑到以后的建筑用途的增加(如加层等)，如图 9.8 所示。

Figure 9.7　Digging of Trench
图 9.7　沟槽开挖

Figure 9.8　Prepare Bottom of Foundation
图 9.8　准备基础底面

### 4. Place the Reinforcement of the Columns for Walls 设置构造柱

Reinforcement bars of columns — previously assembled as a basket — are placed and fixed into the foundation, as shown in Figure 9.9.

柱内钢筋通常绑扎成型嵌固在基础中，如图 9.9 所示。

The basket of hoops must have enough space to let in concrete vibrator device into the column.

钢筋间距必须有足够的空间以确保混凝土振捣棒能插入柱子中，进行振捣。

# Masonry Structure

Figure 9.9  Place the Reinforcement of the Columns for Walls

图 9.9  设置构造柱

### 5. Place the Simple Concrete in The Foundation  在地基中浇筑素混凝土

With reinforcement of all columns placed and provisionally fixed, continuous foundation is filled with simple concrete, as shown in Figure 9.10. For foundation the mix of simple concrete contains a cement-(sand-gravel) ratio of 1:10 plus 30% of big stones. For the over footing the cement-(sand-gravel) ratio for the mix is 1:8 plus 30% of medium stones.

在所有的现浇柱或预制柱和连续基础内都要填充素混凝土，如图 9.10 所示。对基础来说，素混凝土的水泥和砂砾石比例为 1∶10，再加上 30%的大石子。在基础以上，则是以水泥和砂砾石比例为 1∶8 再加上 30%的中石子。

Figure 9.10  Place the Simple Concrete in the Foundation

图 9.10  在地基中浇筑素混凝土

### 6. Detail example of foundation  基础工程实例

Care should be taken when transporting fresh concrete from mixer discharge to the trench, and also in placing concrete in order to not separate stones from fresh concrete. A good curing should be performed, allowing the concrete to reach enough strength, impermeability and durability. Lack of curing causes low resistance and it could appear cracks due to the contraction for drying of concrete. Detail example of foundation is shown in Figure 9.11.

当混凝土从搅拌站运送到沟槽时，应该采取措施进行搅拌使其充分混合。要进行充分的养护以使其达到足够的强度、抗渗性及耐久性。由于混凝土的干缩作用，如果未得到充分的养护可能会引起裂缝的出现。如图 9.11 所示为基础工程实例。

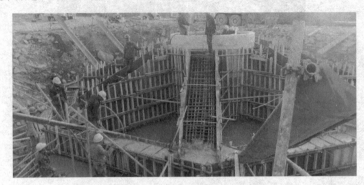

Figure 9.11　Detail Example of Foundation

图 9.11　基础工程实例

## 9.6　How to Build the Over Footing　地梁施工

Over the run foundation continues the over-footing to be used as support for the wall. The main purpose is to isolate the wall from the soil and provides protection against humidity. Here it is used wood boards as a form or mold the over footing, as shown in Figure 9.12.

在基础上继续进行地梁的施工，用于支撑墙体。主要目的是分隔墙体和土壤，且起到防潮的作用。如图 9.12 所示为采用木板作为模具。

Figure 9.12　Build the Over Footing

图 9.12　建造地梁

If soil conditions are not good, like soft sand or flexible soil, the over footing must have minimum steel reinforcement to work as a foundation connection beam.

如果土壤条件不是很好，像软砂或泥泞土壤，则需在垫层内布置钢筋作为基础的地梁。

## 9.7　How to Build A Wall　墙体施工

To build the wall we must prepare the mortar and prepare the bricks before start the process.

## Masonry Structure 09

Above the over-footing, it starts layering of brick units over mortar bed, forming masonry walls.

为了建造墙体我们必须在开始工程前准备好砂浆和砖。在地梁上面开始分层砌筑，设置砂浆层，形成砌体墙。

### 1. Preparation of the Bricks 砖块的制备

Bricks should be wet before laying them so they don't absorb water from mortar and obtaining a good adherence between mortar and brick.

在砌砖之前应把砖浸湿，这样砖块就不会吸收砂浆的水分，砂浆和砖之间形成很好的黏结性。

### 2. Preparation of the Mortar 砂浆的制备

To make the mortar, the mix will have thick sand - cement ratio of 5:1. Sand and cement should be mixed dry. Next this dry mix is put in the tray mixing it with water.

为了制作砂浆，粗砂和水泥按照 5∶1 的比例进行混合，砂和水泥应该一起干燥，随后将干燥的混合物放到托盘上与水一起混合。

### 3. The Construction Process 施工过程

The construction process of wall is shown in Figure 9.13.

墙体施工过程如图 9.13 所示。

Cement Thick Sand     水泥厚砂

Water     水

To make the mortar, sand and cement should be mixed dry, out from the tray. Next, this dry mix is put in the tray mixing them

为了制备砂浆，应先在托盘外干拌砂浆和水泥，然后将干拌好的混合物放入托盘中再次搅拌

Figure 9.13 The Construction Process of Wall

图 9.13 墙体施工过程

Masonry units (bricks)Bricks should be wet before laying them, so they don't absorb water from mortar, and obtaining good adherence mortar-brick

在砌砖之前应把砖浸湿，这样砖块就不会吸收砂浆的水分，砂浆和砖之间形成很好的粘结性

Over mortar bed are placed the bricks

在砂浆层上放置砖

Each layer of brick, should be checked vertically using a lead weight

用铅块检查每层的砖块是否垂直

Using a palette or brick layer's trowel mortar are placed over bricks in a way mortar penetrate bricks holes

用抹子把砂浆放在砖层上，然后用砂浆填充砖隙

Figure 9.13　The Construction Process of Wall(Continued 1)

图 9.13　墙体施工过程(续一)

Bricks at lateral edges of wall will be the "master brick". With the help of a screed board and a cord fixed between master bricks, it is verified horizontal level and the thick of mortar joint, which should be the same for all bricks in the layer. Do not build more than 1-2m height of wall per day

墙的侧边上的砖应该是干燥的砖。在准条板的帮助下，使砖层的厚度都一样，并且砂浆与砖结合点垂直。墙到达了最终的高度，每一层都要检测墙的垂直度，每天砌筑不要超过 1~2m 墙高

Figure 9.13　The Construction Process of Wall(Continued 2)

图 9.13　墙体施工过程(续二)

## 9.8　How to set the Confining Columns to the Wall
## 　　　构造柱施工

Confining columns are shown in Figure 9.14.

构造柱如图 9.14 所示。

Be sure the reinforcement bars of the columns and stirrups were placed and fixed properly at the stage of the foundation.

Maximum distance between confining columns in a 14cm thick wall is 3.50m and for a 24cm thick wall is 5.00m. In both sides of a wall an empty place for column place when layering bricks edges of wall are "teething" (castled vertical edge) making layers not aligned vertically at the edges of wall-as shown in photo, and obtaining a better tying or anchorage between column and wall

首先确保构造柱的钢筋与基础连接完好。若墙厚为 14cm，构造柱之间的最大距离为 3.5m；对于 24cm 的墙，构造柱间距为 5m。在构造柱与墙体间，两边都有孔隙，且砖凹凸相隔，以保证构造柱与墙体间的锚固作用

Figure 9.14　Confining Columns

图 9.14　构造柱

1. Placing the Forms (as shown in Figure 9.15)　模板工程(如图 9.15 所示)

The forms are made from wood panels or steel plates. Bracing of the panels is needed to assure stability of the form.

模板可为木板或者钢板，并保证其有一定的稳定性。

Figure 9.15　Placing the Forms

图 9.15　模板工程

2. Placing of Concrete(as shown in Figure 9.16)　混凝土的浇注(如图 9.16 所示)

The concrete must be transported by the operator in clean cans and dropped from the top of column. The process must continue in order to assure uniformity of the mix and avoid dry joints among it. Vibration of the poured mix is required.

混凝土应采用干净的罐体运输，从柱子的顶部往下浇注。这个过程必须连续以确保充分混合，避免浇注的连接部位失水。浇注的过程中需要振动。

Figure 9.16　Placing of Concrete

图 9.16　混凝土浇注

## 9.9　How to Build the Slab and Beams　梁、板施工

For concrete element (columns, beams, stairway and slab), reinforcement are corrugated bars of steel, cut in appropriate length. After finishing wall construction and with arrangement of beam reinforcement, forms for slab are placed.

对混凝土构件(柱、梁、楼梯和板)中的螺纹钢来说，要截取适当的长度。在完成墙体施工、梁的布置后，楼板的模具就被确定了。

Minimum covering for concrete elements placed at site construction is shown in Table 9.1.
现场施工混凝土最小保护层厚度如表 9.1 所示。

Table 9.1  Minimum Covering for Concrete Elements Placed at Site Construction

表 9.1  现场施工混凝土最小保护层厚度

| Description<br>类型 | e(cm) |
|---|---|
| Elements in contact with ground or exposed to weather<br>接触地面或暴露于天气 | |
| For diameters equal or smaller than $5/80h$<br>直径等于或小于 $5/80h$ | 4 |
| For diameters more than $5/80h$<br>直径超过 $5/80h$ | 5 |
| Elements placed above ground or in contact with sea water<br>直接放置地面或与海水接触的构件 | 7 |
| Elements neither in contact with ground nor exposed to weather<br>既不接触地面也不暴露于空气中的构件 | 2 |
| Light slabs<br>灯板 | 2 |
| Walls or shear walls<br>墙壁或剪力墙 | 4 |
| Beams and columns (measured to the stirrup or hoop)<br>梁、柱(测量到箍筋或箍) | 2 |
| Shells<br>板壳 | |

The placing of concrete starts with the slab beams followed by 5cm depth of concrete cover over the slab. During the placement of concrete for slab, thickness of concrete should be checked. A way to do this leveling is making separated strips of concrete with specified level. This procedure is repeated side by side successively finishing the entire slab.

混凝土的铺设需要在板梁上面先铺设 5cm 厚的混凝土板。在布置混凝土板期间，需要检查一下混凝土的厚度，选择合适的木榫或者木尺做标准测量从砂浆顶部到需要的水平位置，这是一个找平的过程，直到整个板都被填平。

For concrete with strength resistance of $210 kgf/cm^2$, the volume proportioning of materials is 1 of cement, 2 of stone and 2 of sand. The water cement ratio is around 0.45. Amount of water can be varied based on weather conditions, temperature and other external factors. A mixer machine is recommended to use for the mix of the concrete. Ingredients of the mix are input on the machine in the following order, first introduce 1/4 of the water amount, then the stone and the sand, mixing as well as possible, to finish the cement, to complete the 3/4 of water at the end of the mix.

对于强度为210kgf/cm² 的混凝土，其体积配比为水泥∶砂∶石为1∶2∶2，水灰比约为0.45。具体的用水量可根据天气、温度及其他因素略有不同。建议用搅拌机搅拌混凝土。放入搅拌机中的混合材料按以下的顺序：首先放入1/4的水，然后放入石子和砂子，尽可能地搅拌好，最后放入3/4的水，完成搅拌混凝土。

A good vibration process must be produced to avoid voids in the concrete. Vibrators should be used for vibration. If voids or irregularities are appeared in the concrete, the resistance in the concrete will decrease.

正确的振捣使混凝土中没有孔洞。振捣过程中要用到振捣棒。如果混凝土中出现了不规则的孔洞，那么最终的耐久性会下降。

Slab should be cured immediately after concrete start to harden (at initial setting) during 7 days at least. The first days or first night of curing is the most important. Special attention should be paid to thin slabs or any structural element exposed to the weather. Forms of slab can be retired after 7 days after its placement. A covering of flat bricks or a mud covering can be placed above slab in order to prevent the slab getting wet from rain.

在混凝土开始硬化的7天内，需要立即对混凝土板进行加工处理。第一个白天和第一个晚上的处理最重要，对薄板和暴露在空气中的部分要进行特殊处理，浇注7天后可以撤掉模板。为了防止模板被雨淋湿，可以在板上面加些砖或泥来覆盖模板。

After placement of concrete, a wood plate or brick layer's trowel is used to level the surface giving it a better finish.

混凝土浇注后，要用木铲或砖铲对混凝土表面进行平整，使其有一个更好的光洁度。

## 9.10 How to Finish the Surface of the Elements 构件表面做法

For finishing of wall and ceiling surface it is necessary to use scaffolds, so covering works with mortar can be made at the entire height of elements. It is starts from the upper part and then run down to the lower part. The mix in volume has a proportion of 1 of cement to 3 of fine sand.

It is very important to keep the mortar workable, so the proportion of mixture must remain identical during the whole process.

在完成墙壁和天花板表面工程时，需要进行饰面，在构件的整个高度用砂浆覆盖表层，从上往下进行。混合物为1∶3的水泥混合砂浆。

保持砂浆的和易性是非常重要的。所以要求混合砂浆在整个过程中必须保持相同的比例。

Finishing the surface of the elements is shown in Figure 9.17.

表面做法如图9.17所示。

Figure 9.17　Finishing the Surface of the Elements

图 9.17　表面做法

  Finished all the structural work, door and windows installing, wall and ceiling can be painted. At first sanding process over the elements is performed in order to discover irregularities. Then putty process to cover the imperfections must be done prior to the application of the base paint. Finally paint finishing is put over the walls.

  完成所有结构构件的施工及门和窗户的安装，才可进行墙壁和天花板的粉刷。首先要打磨构件，以便及时发现施工不规范之处。填补工程应在粉刷工程之前完成，最后进行墙体等部位的粉刷。

## New Words and Expressions　生词和短语

1. Masonry Structure　　砌体结构；砖石建筑物；石工结构
2. clay　　*n.* [土壤] 黏土；泥土；肉体；似黏土的东西
3. shale　　*n.* [岩] 页岩；泥板岩
4. seismic　　*adj.* 地震的；因地震而引起的
5. mechanical property　　[机] 机械性能；[力] 力学性质
6. workmanship　　*n.* 手艺，工艺；技巧
7. configuration　　*n.* 配置；结构；外形
8. humidity　　*n.* [气象] 湿度；湿气
9. breakable　　*n.* 易碎的东西，易破的东西
     *adj.* 易碎的
10. lime-siliceous brick　　石灰硅质砖
11. hollow　　*n.* 洞；山谷；窟窿
    *adj.* 空的；中空的，空腹的；凹的
12. tubular　　*adj.* 管状的
13. petroleum　　*n.* 石油
14. rope　　*n.* 绳，绳索

15. embedded    v. 嵌入(embed 的过去式和过去分词形式)
              adj. 嵌入式的；植入的；内含的
16. triangle    n. 三角(形)；三角关系；三角形之物
17. theodolite   n. 经纬仪
18. landfill    n. 垃圾填埋地；垃圾堆
19. trench     n. 沟，沟渠；战壕；堑壕
20. gravel     n. 碎石；砂砾
21. vibration   n. 振动；犹豫；心灵感应

# Exercises　练习

Ⅰ. Write a T in front of a statement if it is true according to the text and write an F if it is false.

1. The first masonry structures were unreinforced and intended to support mainly gravity loads.

2. The water of masonry construction shouldn't contain filthy elements, should be clean, drinkable and fresh.

3. It is important that foundation to be leveled below the ground level, on natural soil at a depth not less than 1.2m.

4. It is recommended to use a mixture ratio of cement, sand and gravel for the over footing of 1:8 plus 30% of medium stones.

5. Bricks should be wet before laying them so they don't absorb water from mortar and obtaining a good adherence between mortar and brick.

Ⅱ. Complete the following sentences.

1. Masonry construction has been used in a variety of structures:_____.
2. Please five materials of the masonry construction:_____.
3. Sand will use on the mix with _____.
4. Masonry units can be _____. Solid section without holes must be more than _____ of the geometrical area.
5. For foundation the mix of simple concrete contains a cement-sand-gravel ratio of _____ plus _____ of big stones.
6. For diameters more than 5/80h, the minimum covering for concrete elements placed at site construction is _____.

Ⅲ. Translate the following sentences into Chinese.

1. In the last 45 years, the introduction of engineered reinforced masonry has resulted in structures that are stronger and more stable against lateral loads, such as wind and seismic.

2. Ropes (cord) are tightened, using trestles made by wood poles nailed to a transversal stick

and embedded to the ground.

3. A trench digging for continuous foundation should be made following the structural plans and details.

4. Foundation dimensions should consider future expansion of the building like the increasing of the stories, at the time of the design.

5. To build the wall we must prepare the mortar and prepare the bricks before starting the process.

6. Bricks should be wet before laying them so they don't absorb water from mortar and obtaining a good adherence between mortar and brick.

Ⅳ. Translate the following sentences into English.

1. 第一个砌体结构是无筋的，主要支撑重力荷载。
2. 在秘鲁和南美的房屋，黏土砖砌体建设是最有可能使用的结构类型。
3. 保存条件应干燥、防水；否则它将湿润，膨胀起来，变得柔软。
4. 然而如果未受到控制的垃圾填埋场或垃圾沉积物构成了地面，那么大型居住区有可能建在这些基础上。
5. 主要目的是把墙与土壤隔绝，以为其提供保护进而防止潮湿。
6. 混凝土必须被操作者运输到干净的罐中，然后从柱的顶端倒入。

# Answers　答案

Ⅰ.

1. T　2. T　3. F　4. T　5. T

Ⅱ.

1. homes, private, public buildings, historical monuments
2. cement, sand, stone, gravel, water
3. cement, stone and water
4. solid, hollow or tubular; 75%
5. 1:10; 30%
6. 5cm

Ⅲ.

1. 在过去的45年中，设计配筋砌体的引入使得结构更强，对于横向荷载更为稳定，如风力和地震。
2. 把绳拉紧，利用支架把木杆钉在横棍上，嵌入地面。
3. 对于连续基础的沟槽开挖应遵循结构图和细节。
4. 基础的尺寸应考虑建筑将来的发展，例如在设计时间内楼层的增加。
5. 为了修筑墙体，我们必须在开始之前准备砂浆和砖。
6. 砖在使用之前应该是湿的，这样可使它们不从砂浆吸水，并获得良好的粘结砂浆砖。

Ⅳ.

1. The first masonry structures were unreinforced and intended to support mainly gravity loads.

2. Clay brick masonry building is the most likely used type of structural system on housing in Peru and South America.

3. Forms should be dry and protected from water; otherwise it remains humid, swells up and becomes soft.

4. However, if non-controlled landfill or garbage deposits compose the ground, large settlements are expected on the foundation.

5. The main purpose is to isolate the wall from the soil and provides protection against humidity.

6. The concrete must be transported by the operator in clean cans and dropped from the top of column.

# Chapter 10　Building with Wood
# 第 10 章　木　结　构

This Chapter focuses on both the range of opportunities and the limitations of building with wood.

本章主要介绍木结构的优点和局限性。

It notes, as well, certain advantages over traditional construction systems in China.

同时也指出相比之下中国传统建筑体系的一些优势。

Wood structures are not intended to support high rise buildings, although wood is used as components in tall buildings, including infill walls, roof structures, and other forms of hybrid construction combining concrete and steel structures with wood.

虽然木材可用作高层建筑的某些部件，如填充墙、屋顶，还可以和混凝土一起用作混合建筑，钢结构也会用到木材，但是木结构本身并不能用作高层建筑。

Japan, Europe, Canada and the United States have a long tradition of wood construction, and have developed modern wood building into competitive solutions for the low to mid-rise segment. However, and as noted in the Chapter, China has a much longer tradition of building with wood than North America and most likely even longer than Europe.

日本、欧洲、加拿大和美国都有悠久的木结构建筑历史，在现代从单层到中高层都应用了木结构建筑。然而正如本章所述，中国的木结构建筑历史比北美洲更长，甚至很可能比欧洲更悠久。

This Chapter looks to the immediate as well as longer term possibilities for wood construction in China, including exciting structural forms and the use of engineered wood products such as glulam beams and columns.Here, the architectural beauty and warmth of wood, which so many Chinese have noted, is open to view and on display.

本章着眼于应用木结构建筑在中国目前和未来长期的可行性，这其中包含多样的建筑结构和工程中使用的木材制品，如胶合木梁、胶合木柱。中国人比较在意的建筑的美观、木材的保温，都可以在本章了解到。

Traditional Chinese wood house is shown in Figure 10.1.

传统的中国木房子如图 10.1 所示。

Wood Frame apartment building is shown in Figure 10.2.

木结构公寓楼如图 10.2 所示。

Figure 10.1 Traditional Chinese wood house
图 10.1 传统的中国木房子

Figure 10.2 Wood frame apartment building
图 10.2 木结构公寓楼

## 10.1 Wood Frame Construction: Low-rise Solutions
木框架结构：低层建筑

Wood frame construction is already being used for housing in China, from single family dwellings in the suburbs of cities such as Shanghai and Beijing, to low-cost rural developments where land availability is not a problem. They have proved cost-competitive and perform well in comparison with concrete and steel frame housing.

木结构已经被用于中国的低层房屋，从北京和上海郊区的独栋家庭住宅到土地供应不成问题的低成本农村住宅。这些建筑证明了木结构的高性价比，也展示了木结构相较混凝土结构和钢结构房屋的优越性。

Semi-detached wood frame house in Sichuan is shown in Figure 10.3.
四川半独立式木结构住宅如图 10.3 所示。

Figure 10.3 Semi-detached Wood Frame House, Sichuan
图 10.3 四川半独立式木结构住宅

But much more can be done. Wood construction is the solution to other building requirements in China as well. These include medium-density multi-storey apartments, small commercial and office buildings, schools, medical clinics, nursing homes, universities and research centres, sports arenas and other recreational facilities.

但是木结构还有很大的潜力，木结构也可以满足其他建筑的需求，如中等密度的多层公寓、小型的商业楼和办公楼、学校、诊所、养老院、大学、研究中心、运动场和很多娱乐设施。

Wood frame construction: medium-rise solutions
木框架结构：多层建筑

Wood can make a contribution to solving China's housing shortages through high density multi-family solutions. While these can take the form of two or three-storey apartment blocks, the future lies in the higher-rise buildings which are well-proven in Europe and North America.

通过高密度多住户的解决方案，木结构可以为解决中国的住房短缺问题做出贡献。目前这些两层或三层的公寓楼，会逐渐向欧洲和北美的高层结构发展。

They have gained popularity in these regions because of lower building costs, wood's suitability for highly efficient industrial building methods, better energy-efficiency, better seismic performance and a growing environmental awareness. And, because of their low weight, multistorey wood buildings can be constructed without the need for extensive pile foundations. This makes it possible to develop sites which would previously have been impractical.

由于建筑成本低廉，在这些地区木结构非常受欢迎，木材能够完美地适应高效工业建筑方法，有很好的节能性、抗震性、环保性。并且，由于木材比较轻，多层的木结构房屋也不需要设置很多桩基础，这使得在许多以前不适合开发的地方进行建筑变为可能。

In China, as of 2009, existing fire codes do not allow wood frame apartment blocks of four or more storeys. However, this may be an option for the future, as these codes are often under review and the scientific experience supports more storeys.

2009年，中国出台了防火规范，其中规定木结构的公寓最高为四层。然而，这可能是未来的一种选择，因为规范也是需要检验的，研究人员正在试验更高层的木结构建筑。

5-storey wood frame apartment building in Europe is shown in Figure 10.4.
欧洲五层木结构建筑如图10.4所示。

Figure 10.4   5-storey Wood Frame Apartment Building, Europe
图10.4   欧洲五层木结构建筑

## 10.2 Wood Frame in China 木结构在中国的发展

At present, wood frame is used for single family and multi-family homes of two or three storeys in China, as shown in Figure 10.5. Wood members form a structural framework which is sheathed with structural wood panels. Foundations are generally concrete. The floor above can be either wood or a concrete slab and forms the platform for the next storey. Roof and wall insulation and water-proof membranes provide energy-efficiency and protection from moisture. Interiors are usually dry-lined with fire-resistant gypsum board, and many different materials can be used for external cladding. Because the structure has multiple wood members, panels, fasteners and connectors, loads can be carried through a number of alternative pathways. As a result, wood frame buildings are highly resistant to sudden failure in earthquakes or high winds.

目前在中国木框架结构只有两层或三层，用做一户或多住户的住宅，如图 10.5 所示。木结构的骨架是由木板制成的，基础通常是混凝土结构。地上的楼板可以是木板也可以是混凝土板。屋顶、墙体保温、屋面防水不仅使能源高效利用，也让建筑避免潮湿。内部是干燥的耐火石膏板，外层的涂料是多种多样的。木结构中有许多由木材制成的构件，如面板、紧固件、连接件，都有很高的承载能力。总之，木结构房屋对于突然的地震、大风都有很强的抵抗能力。

Figure 10.5  Multiple Living-unit Wood Frame Construction, Sichuan
图 10.5  四川多层木结构建筑

Multi-family units are built using the same techniques and under the same building codes as single family units, and are separated by code required fire-rated assemblies. The units generally range in size from 100 to 300 square meters.

多家庭的木结构建筑与一户的木结构遵循同样的建筑规范，由规范所要求的防火组件分隔开来。多住户木结构的建筑面积在 100～300 平方米之间。

Multi-storey wood frame buildings are popular in many countries. Where five and six-storey blocks are now being built, apartments are generally on a single storey, separated from each other by fire resistant assemblies. Horizontal stability in these taller buildings is achieved using engineering design which incorporates braced walls and heavy-duty metal connections between assemblies. Noise is an important consideration, too. Effective solutions are available to limit sound transmission through floors and walls.

多层的木结构建筑在许多国家都非常流行，五层和六层的公寓楼现在正在建造中，每层的套间都由耐火组件隔开。在这些高层建筑中，工程设计通常使用坚固的承重墙和耐用的金属连接件来保证每一层的坚固。噪声也是需要考虑的问题，地板和墙可以很好地防止声音传播。

## 10.3 Hybrid Construction: Wood Frame Storeys on Concrete Structure 混合建筑：木框架与混凝土结构结合

Hybrid construction, where wood construction is combined with concrete and/or steel, is a promising opportunity for the future of China. This includes the construction of buildings which have the lower storey or storeys (or parkade) in concrete, to which a light-weight energy-efficient wood super-structure can be attached.

混合结构，指木结构与混凝土和钢筋结合，在中国很有发展前景。这些结构包括低层混凝土结构或单层混凝土结构(停车拱廊)，它们同轻质节能的木结构相连接。

In Europe and North America, wood frame buildings of up to six or seven storeys are achieved using a concrete lower storey. In China, buildings of up to three wood frame storeys on top of up to four concrete storeys may soon be accepted.

在欧洲和北美洲，六层或七层的木结构建筑在底层会用混凝土结构。中国可能很快就会认可低层四层混凝土结构上叠加三层木结构的建筑形式。

These hybrids can combine commercial space, such as stores and offices, in the concrete portion of the building, with housing in the wood frame part, as shown in Figure 10.6. In some settings, hybrids may be the most practical, efficient, and cost-effective option.

这种混合形式可以用在商业，如商店和办公室，在建筑物混凝土的部分建造木结构住宅，如图 10.6 所示。从某种程度上说，混合结构是一个非常实用、高效、低成本的选择。

Figure 10.6 Multi-family Wood Frame Apartment Building on Concrete Parkade, Canada
图 10.6 加拿大建在混凝土停车场上的多住户木结构公寓

1. Hybrid Construction: Wood Frame Walls in Concrete Structures (as shown in Figure 10.7)  混合建筑：在混凝土建筑上的木框架墙(如图 10.7 所示)

Figure 10.7  Assembling Pre-fabricated Wood Frame Exterior Infill Wall Panels in a Multi-storey Concrete Structure Building, Europe

图 10.7  欧洲多层混凝土结构上装配预制外木框架填充墙

  Chinese fire safety codes allow the use of infill wood frame exterior walls in concrete structures up to six storeys, soon likely to be extended to seven storeys high for residential, offices, and certain factories and warehouses. These structures have been built cost-competitively at up to twenty storeys in northern Europe for a number of years, where increasingly stringent energy-efficiency requirements are a key driver.

  中国消防安全规范允许在混凝土结构使用最高六层的木框架外部填充墙，很快住宅、办公楼、一些工厂和仓库可能会扩展为 7 层。多年来在北欧这种类似的建筑已经可以建到 20 层，具有很强的成本竞争力，越来越高的能效要求是关键驱动力。

  Exterior infill walls are light, as they are designed to take only the load of their weight and the wind and seismic loads that directly affect them. They can be pre-fabricated in a factory or built on-site and have very good insulation characteristics in relation to their thickness, providing substantially better energy performance than traditional concrete, masonry or steel construction.

  外部填充墙非常轻，它们设计为只承受自身的重量，风、地震都会对其有直接的影响。填充墙可以提前预制，也可以在现场制成。由于厚度的原因，填充墙有很好的绝缘能力，并且比传统的混凝土结构、砌体结构和钢结构更加环保。

  Where wood frame is used for interior walls in concrete and steel structures as partitions, it provides flexibility of design, including floor layout, fire safety, sound insulation and renovation. Wood infill partitions are non-structural, lightweight, and are suitable for a range of interior finishes. They can also be designed to meet the fire and sound requirements for apartment partition walls. Wood frame partitions are approved up to eighteen stories.

  木框架不仅可以用于混凝土结构和钢结构的内墙作为隔板，还具有设计的灵活性，如用作地板、防火、隔音和装修。木质的填充物不能用于结构承载，因其非常轻，故而更适合做内部的装修，也可以按需要设计成防火墙和隔音墙。木框架部分最高可以达到 18 层。

## 2. Hybrid Construction: Wood Frame Roofs on Concrete Structures
混合结构：混凝土结构的木框架屋面

Many of the typically concrete medium-rise residential buildings throughout China have flat roofs that tend to leak and are poorly insulated for energy conservation and thermal comfort. These existing roofs can be covered with a pitched wood frame truss roof, as shown in Figure 10.8. This is a cost-effective way of keeping the rain out, improving the look of the building and, with additional insulation in the roof cavity, reducing energy costs. It is also an effective way of delivering a thermally comfortable attic space for extra accommodation, or of installing mechanical systems for heating, cooling, and ventilation.

在中国，许多典型的中型住宅建筑都有平屋顶，屋顶往往会漏水，保温隔热性很差，也不利于能源节约。现存的这种屋顶可以覆盖一个斜的木框架屋顶，如图10.8所示。这种成本低廉的方式既美观又能防雨，减少了资源损耗，也提供了一个温暖的可以居住的阁楼，或者用来安装可加热、降温、通风的机械设备。

Figure 10.8　Re-roofed Apartment Buildings with Habitable Attic, Downtown Beijing

图10.8　北京市中心重做屋顶的带阁楼的公寓楼

This system is as competitive for installing roof systems on new concrete structures as for replacing old concrete roof systems.

这种在新的混凝土结构上安装木框架屋面的系统与替换原有的混凝土屋面系统相比，同样具有竞争力。

## 10.4　Engineered Wood Construction: Solid Wood Panels
工程木结构：实木板

Solid wood panel structures provide a leading-edge alternative for six to ten-storey buildings.

Although the technology is relatively new and not yet recognized in Chinese codes, it is widely used across Europe (as shown in Figure 10.9). The tallest built so far is a nine-storey residential building in London, England.

坚硬的木板结构对于 6~10 层建筑非常适用。尽管这种新技术目前还没有被中国的各项建筑规范认可，但是它已经被广泛地应用于欧洲(如图 10.9 所示)，迄今为止最高的木板结构建筑是英国伦敦的一所住宅楼，高达 9 层。

Figure 10.9　Assembling Solid Wood Panels in Multi-storey Apartment Building, Europe
图 10.9　欧洲多层公寓的装配式实木板

Cross-laminated boards are glued together and used to build walls and joists. Panels are machined in a factory to fine tolerances by computer-controlled equipment. The panels arrive on site with apertures for doors and windows, and wiring and plumbing channels already prepared. The walls can be insulated to provide a high level of energy-efficiency. Superior load-carrying characteristics, including lateral stability against wind and seismic forces, as well as excellent fire safety performance, make cross laminated timbers suitable for medium and even high-rise buildings. And the amount of timber used means buildings made with solid wood panels are highly effective carbon stores.

层压板内部用胶连接，用做墙和托梁。板在工厂由精密的计算机控制的设备加工而成。在到达现场时，木板预留门窗的孔、电线和水管管道也布置好了。墙可以绝缘，提高了能源的利用效率。较高的横向稳定性可以抵挡风和地震，防火能力也很强，这些优良的特性使得层压板非常适合中高层建筑。使用的木材数量意味着建筑是由实木板制成的，这是非常有效的碳储存。

These environmentally-friendly solid wood buildings offer longer-term opportunities in China, particularly for high density housing requirements.

这些有利于环境保护的木结构建筑尤其可以满足高密度的房屋需求，在中国有非常长远的发展前景。

## 10.5 Engineered Wood Construction: Glued Laminated Timber
工程木结构：木料胶合板

In North America and Europe, structural glue laminated timber is widely used in constructions where span width is an issue and/or the unique beauty of the wood is to be exploited architecturally. Glulam beams and columns have a strong aesthetic appeal, as the structure of a building can be expressed in the exposed beauty of the wood, as shown in Figure 10.10.

在北美洲和欧洲，胶合板结构广泛应用于大跨度和对外形美观要求高的建筑。胶合板梁、胶合板柱能够满足较高的美学要求，使得木结构建筑的外形非常漂亮，如图 10.10 所示。

Figure 10.10  New Temple with Glulam Post and Beam Structure, Zhejiang
图 10.10  浙江省胶合木板新型梁柱式寺庙

Glue laminated timber-engineered wood beams and columns are used in homes, schools, sports halls, railway stations, industrial and commercial buildings, such as shopping centres and expo buildings, and public buildings, such as museums and concert halls. They are also used in landscaping and infrastructure applications such as glulam bridges.

胶合板梁、胶合板柱被用于住宅、学校、运动馆、火车站、工业及商业建筑，如购物中心、展览馆，还可以用于公共建筑，如博物馆、音乐厅。胶合板也可以用于美化装饰和一些基础设施，如胶合木桥。

Glulam beams and columns come as standard products, with a variety of cross-sections and lengths. Custom designed beams and columns are pre-fabricated according to customer needs and can include curved shapes and mechanical interfaces to concrete or steel structures etc.

作为标准产品的胶合板梁柱，具有多种截面和长度。可以根据顾客的需求定制梁和柱甚至是曲形构件，也包括混凝土结构和钢结构的连接件等。

Glulam is a mature technology in Europe, where large span buildings are still in use after almost 100 years. Modern design methods are available and national codes-design, production, fire-supporting glulam construction in China was approved in 2010.

在欧洲，胶合板的技术非常成熟，很多大跨度的建筑在 100 年后仍然可以使用。胶合板的现代设计方法是可行的，2010 年中国批准了胶合板国家标准设计、生产，以及防火胶合板建筑。

**Wood Construction: On-site or Prefabrication　木结构：现场制作或预制**

The traditional way to construct wood frame buildings in North America is on-site, particularly when there is labour availability. In Europe, wood frame assemblies are typically pre-fabricated. Engineered wood construction, such as glulam, is most commonly erected piece by piece on-site. In China, almost all wood buildings are currently constructed on-site, as shown in Figure 10.11. Building materials and structural components are freighted to the building site and the various assemblies — walls, floors, etc. — are framed on-site. The method requires organization and planning on the building site and measures must be taken to avoid moisture damage to materials. On-site construction relies on a skilled work force and, while much faster than using other materials, is slower than using pre-fabricated elements.

北美在劳动力允许的情况下，建筑木结构房屋的传统方法是在现场施工。在欧洲，木框架的部件是提前准备好的，工程中的木结构，如胶合板，通常就是一片一片地立在现场。在中国，几乎所有的木结构建筑现在都是现场施工，如图 10.11 所示。建筑材料和结构部件被运输到现场，各种各样的装配墙、地板也是在现场组装的。采取这种方式要合理安排施工现场，并且必须通过一些技术措施避免建筑现场的水分腐蚀材料。现场施工需要熟练的技术劳动力，虽然木结构现场组装的方式比使用其他材料更快，但是还是比使用预制木构件慢。

Figure 10.11　On-site Construction of New Roof on Existing Apartment Building, Beijing
图 10.11　北京现有公寓楼现场搭建新屋顶

On-site construction does not require the initial capital costs for plant and machinery, nor the need to maintain capacity utilization. It is particularly appropriate where housing volumes are not large, where labour is reasonably priced and plentiful, and where flexibility and low overheads are important.

现场施工的方式并不需要一开始就在机械与设备上投入大量的资金，也不需要保持产能利用率，在建筑体积较小、劳动力充足并且价格合理的情况下尤其合适，但是要注意灵活性和低开销。

While more capital-intensive, off-site pre-fabrication has the benefit of controlled factory conditions, less dependence on on-site labour and faster construction times.

资本密集型和异地预制对于工厂加工控制、减轻现场压力以及缩短施工周期非常有利。

In the case of wood frame construction, only a few days on the building site are needed to assemble a water-tight structure, complete with roof. The panels can be pre-fabricated with insulation, windows and doors. Entire units can even be made complete with electricity, water and waste pipes, kitchens and wet rooms, floors and papered walls.

木结构施工只需要短短几天的时间就可以完成整个建筑。预制板、门窗提前进行保温处理。整个建筑设施非常完备，甚至通电、通水、排水、厨房、卫生间、地板和纸墙都可以全部建好。

Pre-fabricated components are relatively light and can be erected at heights of several storeys using simple lifting equipment, such as the cranes on the trucks that deliver components to site. Components may need protection against the elements to prevent dampness.

预制组件相对较轻，可以用简单的升降装置在几层楼的高度上安装，起重机上的卡车就可以为现场运送组件。这些构件需要经过处理来防止现场的湿气。

The extent of pre-fabrication varies widely between countries and companies, depending on economic factors. It does require an up-front investment in plant and equipment which could impose an uncompetitive burden. While this is essentially the case in China at present, over the longer term, pre-fabrication may prove advantageous.

根据经济因素，组件预制的程度在国家和公司之间有很大的不同。预制需要在工厂和设备上有很大的前期投资，这是不利于竞争的一项负担。这基本符合中国目前的情况，可能在长期发展过程中能够证明预制生产是很有利的。

## New Words and Expressions　生词和短语

1. construction　　*n.* 建造；建筑物
2. structural　　*adj.* 结构(上)的，构架(上)的
3. timber　　*n.* 木材，木料；*vt.* 用木材建造
4. adequate　　*adj.* 足够的
5. dimension　　*n.* 尺寸，面积
　　　　　　　*vt.* 把……刨成(或削成)所需尺寸；标出尺寸
6. vertical load　　竖向荷载
7. horizontal forces　　水平力
8. reinforce　　*vt.* 加固
9. columns　　*n.* 柱(column 的名词复数)
10. manufacture　　*vt.* 加工；制造

　　　　　　　　　　*n.* 制造；制成品，产品
11. rhomboid　　*n.* 菱形
　　　　　　　　*adj.* 菱形的
12. triangle　　*n.* 三角(形)，三角板
13. rectangular　　*adj.* 矩形的，直角的
14. compression　　*n.* 压缩，压抑
15. log　　*n.* 圆木，原木
　　　　　　*vt.* 伐(林木)，采伐
16. trunk　　*n.* 树干
17. facade　　*n.* (建筑物的)正面，表面
18. perforated　　*adj.* 穿孔的；有排孔的
19. panel　　*n.* 嵌板
　　　　　　　*vt.* (用镶板等)镶嵌(门，墙等)
20. mortise　　*n.* 榫眼
　　　　　　　*vt.* 榫接；使牢固相接
21. tenon　　*n.* 榫，凸榫，榫舌
　　　　　　*vt.* 在……上制榫，用榫接合
22. component　　*n.* 构成要素，零件
　　　　　　　　*adj.* 组成的；构成的
23. thermal insulation　　热绝缘，保温层
24. manual craft　　手工工艺
25. aesthetic　　*adj.* 美的；艺术的
　　　　　　　　*n.* 美学
26. platform　　*n.* 平台
　　　　　　　*vt.* 把……放在台上
27. erected　　*adj.* 直立的，垂直的
　　　　　　　*vt.* 使竖立，建立
28. skeleton　　*n.* 骨骼 (房屋等的)骨架
29. crosswise　　*adv.* 斜地，交叉地
30. texture　　*n.* 质地；纹理
31. flammable　　*n.* 易燃物
　　　　　　　　*adj.* 易燃的
32. junction　　*n.* 连接；接合点；交叉点
33. coarse　　*adj.* 粗的；粗糙的
34. frame　　*n.* 框架；结构
　　　　　　*adj.* 有木架的；有构架的
35. joist　　*n.* 托梁
36. glulam　　*n.* [木] 胶合层木
37. purlin　　*n.* 檩；桁条

# Exercises 练习

Ⅰ. Write a T in front of a statement if it is true according to the text and write an F if it is false.

1. The water evaporated must be replaced by sufficient quantities of fresh supplies transported to the shoots by an equivalent root system.

2. Cellulose is made up of sugar molecules that fit together in regular manner with their long-chain direction parallel to each other.

3. The wood cells are orientated and glued together by a matrix of lignin.

4. Coniferous wood consists of 85%~90% tracheids, which are prearranged in radial arrays, and their longitudinal direction is oriented along the axis of the stem of the tree.

5. In juvenile wood, the wood cells are relatively short and thin walled with a remarkable slope of the micro fibril or the s2 layer of the secondary wall.

Ⅱ. Complete the following sentences.

1. Wood have gained popularity in Europe and North America because of _____ _____, _____, _____, _____and_____.

2. Because the structure has_____,_____,_____and _____ , loads can be carried through a number of alternative pathways. As a result, wood frame buildings are highly resistant to sudden failure in earthquakes or high winds.

3. In Europe and North America, wood frame buildings of up to _____ are achieved using a concrete lower storey. And in China, buildings of up to _____ _____ on top of up to four concrete storeys may soon be accepted.

4. Solid wood panel structures provide a leading-edge alternative for six to ten-storey buildings. Although the technology is relatively new and not yet recognized in Chinese codes, it is widely used across Europe. The tallest built so far is _____ .

5. Chinese fire safety codes allow the use of infill wood frame _____ , soon likely to be extended to seven storeys high for _____ , _____ , and _____ and _____ .

Ⅲ. Translate the following sentences into Chinese.

1. A tree can be seen as a structure that faces an ongoing optimisation process over millenniums.

2. The water evaporated must be replaced by sufficient quantities of fresh supplies transported to the shoots by an equivalent root system.

3. Roof and wall insulation and water-proof membranes provide energy-efficiency and protection from moisture.

4. Wood frame buildings are highly resistant to sudden failure in earthquakes or high winds.

5. In some settings hydrids may be the most practical, efficient and cost-effective option.

Ⅳ. Translate the following sentences into English.

1. 胶合板：用三层或多层的木材薄片胶结在一起制成的复合板材。
2. 胶合板梁、胶合板柱能够满足较高的美学要求，使得木结构建筑的外形非常漂亮。
3. 资本密集型和异地预制对于工厂加工控制，减轻现场压力以及缩短施工周期非常有利。
4. 木结构施工只需要短短几天的时间就可以完成整个建筑。
5. 预制需要很大的前期投资，组装期间则投资较少。

# Answers　答案

Ⅰ.

1. T  2. T  3. F  4. F  5. T

Ⅱ.

1. lower building costs, wood's suitability for highly efficient industrial building methods, better energy-efficiency, better seismic performance, a growing environmental awareness

2. multiple wood members, panels, fasteners, connectors

3. six or seven storeys, three wood frame storeys

4. a nine-storey residential building in London, England

5. exterior walls in concrete structures up to six storeys, residential, offices, certain factories，warehouses

Ⅲ.

1. 一棵树可以被看作一个面临着上千年来正在进行优化过程的结构。
2. 蒸发的水必须被足够数量的新的补给品取代，这些补给品由根部运输到枝条。
3. 屋顶、墙体保温、屋面防水不仅使能源高效利用，也让建筑避免潮湿。
4. 木结构房屋对于突然的地震、大风都有很强的抵抗能力。
5. 从某种程度上来说，混合结构是一个非常实用、高效，而且低成本的选择。

Ⅳ.

1. Plywood: manufactured panel made up of three or more thin plies (layers) of wood.

2. Glulam beams and columns have a strong aesthetic appeal, as the structure of a building can be expressed in the exposed beauty of the wood.

3. While more capital-intensive, off-site pre-fabrication has the benefit of controlled factory conditions, less dependence on on-site labour and faster construction times.

4. In the case of wood frame construction, only a few days on the building site are needed to assemble a water-tight structure, complete with roof.

5. It does require an up-front investment in plant and equipment which could impose an uncompetitive burden.

# Chapter 11　Composite Structures
# 第 11 章　组 合 结 构

## 11.1　Applications of Concrete-filled Steel Tubes
　　钢管混凝土的应用

　　Using steel and concrete together utilizes the beneficial material properties of both elements. Reinforced Concrete (RC) sections are one example of this composite construction. This type of section primarily involves the use of a concrete section which is reinforced with steel rods in the tension regions.

　　将钢材和混凝土组合在一起能够很好地发挥这两种材料的力学性能。钢筋混凝土截面实际上就是组合结构。这种截面主要是在受拉区布置钢筋来提高截面承载力。

　　This chapter deals with another type of concrete—steel composite construction, namely concrete-filled steel tubes (CFSTs). The hollow steel tubes can be fabricated by welding steel plates together or by hot-rolled process, or by cold-formed process. Figure 11.1 shows some typical CFST section shapes commonly in practice, namely circular, square and rectangular. They are often called concrete-filled CHS (circular hollow section), SHS (square hollow section) and RHS (rectangular hollow section), respectively.

　　本章主要介绍另一种类型的钢——混凝土组合构件，即钢管混凝土。通过由钢板焊接或通过热轧、冷加工工艺制作中空的钢管。图 11.1 给出了在实际工程中常用的典型的钢管截面形状，如圆形、方形和矩形截面，可分别称之为圆钢管混凝土、方钢管混凝土和矩形钢管混凝土构件。

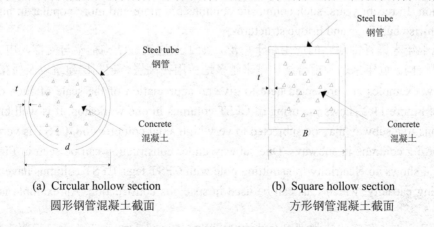

Figure 11.1　Typical CFST Sections
图 11.1　典型的钢管混凝土截面

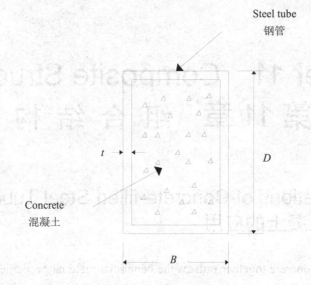

(c) Rectangular hollow section without rounded corners
矩形钢管混凝土截面

Figure 11.1  Typical CFST Sections(Continued)
图 11.1  典型的钢管混凝土截面(续)

In Figure 11.1, $d$ is the outer diameter of the circular section, $B$ is the width of the square or the rectangular sections, $D$ is the overall depth of the rectangular section and t is the steel wall thickness. SHS can be treated as a special case of RHS when $D$ equals $B$. For cold-formed RHS, rounded corners exist as shown in Figure 11.1(c), where r is the external corner radius.

图 11.1 中，$d$ 为圆钢管截面的外径，$B$ 为方形或矩形截面宽度，$D$ 为矩形钢管截面高度，$t$ 为钢管壁厚。当 $D$ 与 $B$ 相等时，则方形钢管混凝土可视为矩形钢管混凝土。对于如图 11.1(c) 所示角部存在倒角的冷加工矩形截面钢管，$r$ 为倒角外部的半径。

There is an increasing trend in using concrete-filled steel tubes in recent decades, such as in industrial buildings, structural frames and supports, electricity transmitting poles and spatial construction. In recent years, such composite columns are more and more popular in high-rise or super-high-rise buildings and bridge structures.

近几十年来钢管混凝土的应用趋于广泛，如工业建筑、结构框架与支撑、电力输送塔架和空间结构。近年来这种组合结构越来越多地应用在高层、超高层建筑以及桥梁结构中。

A few examples are presented here to give an appreciation of the scale of such composite structures. Figure 11.2 shows the using of CFST columns in one workshop. It is well known that the columns in a subway may be subjected to very high axial compression. CFST is very suitable for supporting columns in subways. One subway under construction can be seen in Figure 11.3. Figure 11.4 shows an electricity transmitting pole with CFST legs. CFST columns have very high load-bearing capacity, which thus can be used in spacious construction. An example is given in Figure 11.5.

下面几个例子可表明该类型组合结构的应用前景。图 11.2 所示为钢管混凝土柱在某车间中的应用。众所周知，地铁结构中的柱可能承受很大的轴向压力，而钢管混凝土柱卓越的承压能力非常适合用于地铁结构中，图 11.3 所示是某在建地铁工程。图 11.4 所示为电力传输塔的钢管混凝土支架。由于钢管混凝土具有很高的承载能力，故其广泛地应用于各种空间结构中，如图 11.5 所示。

Figure 11.2　CFST Used in a Workshop
图 11.2　钢管混凝土柱在工业建筑中的应用

Figure 11.3　A Subway Station Using CFST Columns
图 11.3　地铁结构中的钢管混凝土柱

Figure 11.4　A Transmitting Pole with CFST Legs
图 11.4　电力输送塔中的钢管混凝土柱

Figure 11.6 shows the SEG Plaza in Shenzhen during construction. It is the tallest building in China using CFST columns. SEG Plaza is a 76-storey Grade A office block with a four-level basement, each basement floor having an area of $9,653m^2$. The main structure is 291.6m high with an additional roof feature giving a total height of 361m. The steel parts of the columns were shipped to the site in lengths of three storeys. After being mounted, they were connected to the I-beams by bolts and were brought into the exact position. Then, the steel tubes were filled with concrete, and the deck floors were constructed at the same time. In this way, up to two-and-a-half storeys could be built each week, demonstrating the efficiency of this technology. The diameter of the columns used in the building ranges from 900mm to 1,600mm. Concrete was poured in from the top of the column. The concrete was vibrated to ensure the compaction. The SEG Plaza was the first application of circular concrete-filled steel tubes in super-high-rise buildings on such a large scale in China. This technology offers numerous new possibilities, such as new types of

CFST column to steel beam connections, increased fire performance of CFST columns, etc.

图 11.6 所示为建设过程中的深圳赛格广场。这是中国用钢管混凝土柱建成的最高的建筑。赛格广场是一座具有四层地下室、地上 76 层的 A 级办公大厦，地下室每层面积为 9653 平方米。该建筑的主体结构高为 291.6 米，附加的屋顶功能结构层将高度增加到总高度 361 米。钢管部分按三层楼高的长度为一单元运至施工现场。安装完毕后，通过螺栓与 I 字型钢连接至其精确位置。再于钢管中浇注混凝土，同时进行楼面板的施工。以此方式，每星期可以完成两层半的结构施工，证明了这种技术具有很高的施工效率。该建筑中使用的圆钢管直径为 90～1600mm。混凝土从钢管柱顶部开始浇注，辅以振动确保混凝土浇注密实。赛格广场是国内在超高层建筑中第一次如此大规模地应用圆形钢管混凝土柱。这个项目也产生了很多新的施工工艺，如新型钢管混凝土柱与钢梁连接方式、提高钢管混凝土柱的耐火性能等。

(a) During construction
在建过程

(b) After construction
工程完工

Figure 11.5  CFST in Spacious Construction
图 11.5  钢管混凝土在空间结构中的应用

(a)

(b)

(c)

Figure 11.6  SEG Plaza under Construction
图 11.6  建设中的深圳赛格广场

(d)　　　　　　　　　　　　(e)

Figure 11.6　SEG Plaza under Construction(Continued)
图 11.6　建设中的深圳赛格广场(续)

In recent years, CFST columns with square and rectangular sections are also becoming popular in high-rise buildings. Figure 11.7 presents a high-rise building during construction using square and rectangular CFST columns, i.e. Wuhan International Securities Building (WISB) in Wuhan, China. The main structure is 249.2m high, and was completed in 2004.

近年来，方形和矩形截面的钢管混凝土柱在高层建筑中也获得了广泛的应用。图 11.7 所示为武汉国际证券大厦(WISB)，在施工过程中采用了方形和矩形钢管混凝土柱。该建筑的主体结构高度为 249.2 米，于 2004 年完工。

The use of CFST in arch bridges reasonably exploits the advantages of such kind of structures. An important advantage of using CFST in arch bridges is that, during the stage of erection, the hollow steel tubes can serve as the formwork for casting the concrete, which can reduce construction cost. Furthermore, the composite arch can be erected without the aid of a temporary bridge due to the good stability of the steel tubular structure. The steel tubes can be filled with concrete to convert the system into a composite structure and capable of bearing the service load. Since the weight of the hollow steel tubes is comparatively small, relatively simple construction technology can be used for the erection. The popular methods being used include cantilever launching methods, and either horizontal or vertical "swing" methods, whereby each half-arch can be rotated horizontally into position.

钢管混凝土拱桥合理利用了这种结构的优点。在拱桥结构施工中应用钢管混凝土技术的一个重要优点是，中空的钢管可作为模板来浇注混凝土，从而减少工程造价。此外，由于钢管良好的稳定性，钢管混凝土组合拱可以在不借助于一个临时桥的情况下建造起来。在钢管内浇注混凝土形成组合结构后能够承受桥梁在使用过程中出现的荷载作用。由于中空的钢管重量相对较小，因此安装时可采用相对简单的施工技术。该类常用的桥梁施工方法，如悬臂施工法和水平或垂直的"摆动"施工法，都可使每个半拱结构水平旋转至既定位置。

Figure 11.8 illustrates the process of an arch rib during construction. An elevation of the bridge after construction is shown in Figure 11.9. More than 100 bridges of this type have been constructed so far in China. There is much attention being paid both by researchers and the practicing engineers to this kind of composite bridge.

图 11.8 所示为拱肋施工过程。图 11.9 为桥梁施工完成后的立面图。目前为止在中国已建成 100 多座该类桥梁。研究人员和工程师们都对这种组合结构的桥梁予以极大关注。

(a)　　　　　　　　(b)　　　　　　　　(c)

Figure 11.7　Wuhan International Securities Building under Construction

图 11.7　建设中的武汉国际大厦

(a)　　　　　(b)　　　　　(c)　　　　　(d)

Figure 11.8　Elevations of the Arch Rib during Construction

图 11.8　建设中的钢管混凝土拱桥

Figure 11.9　Elevation of the Arch after Being Constructed

图 11.9　建成后的钢管混凝土拱桥

## 11.2　Advantages of Concrete-filled Steel Tubes
　　　钢管混凝土的优点

It is well known that tubular sections have many advantages over conventional open sections, such as excellent strength properties (compression, bending and torsion), lower drag coefficients, less painting area, aesthetic merits and potential of void filling.

众所周知，管状截面具有许多优于传统的开口截面的优点，如出色的强度性能(抗压、抗弯和扭转性能)、较小的阻力特性、无过多的加工部位以及内在的美学特点和潜在的内部填充便利性等等。

Concrete-filled tubes involve the use of a steel tube that is then filled with concrete. This type of column has the advantage over other steel concrete composite columns, that during construction the steel tube provides permanent formwork to the concrete. The steel tube can also support a considerable amount of construction loads prior to the pumping of wet concrete, which results in quick and efficient construction. The steel tube provides confinement to the concrete core while the infill of concrete delays or eliminates local buckling of steel tubes. Compared with unfilled tubes, concrete-filled tubes demonstrate increased load-carrying capacity, ductility, and energy absorption during earthquakes as well as increased fire resistance.

钢管混凝土是在钢管中浇注混凝土。这种柱优于其他类型的钢-混凝土组合柱。在施工中，钢管可作为混凝土的永久性模板。在泵送混凝土前，钢管可以承受相当大的施工荷载，从而使施工过程快速、高效。同时，钢管对核心区混凝土有一定的约束作用，而内部的混凝土会延迟或避免钢管局部失稳。与中空的钢管相比，钢管混凝土构件具有很高的承载能力、延性性能、较好的地震耗能能力及耐火性能。

A simple comparison is given in Figure 11.10 (a) for a column with an effective buckling length of 5m, mass of steel section of 60kg/m and concrete core strength of 40MPa. It can be seen from Figure 11.10(a) that the compression capacity increases significantly due to concrete-filling. A series of tests were performed on void-filled RHS subjected to pure bending. The increase in rotation angles at the ultimate moment due to the void filling was found to be 300%, as shown in Figure 11.10 (b).

图 11.10(a)中给出了简单的对比，一有效屈曲长度为 5m、质量为 60kg/m 的型钢和核心混凝土强度为 40MPa 的组合柱。从图 11.10(a)可以看出，由于混凝土的填充，抗压能力显著增加。研究人员对矩形钢管进行了一系列纯弯曲试验。由于空隙填充，在极限抗弯荷载作用下的截面扭转角变形达到 300%，如图 11.10(b)所示。

A schematic view of interaction diagrams for beam-columns is shown in Figure 11.10(c). It is clear that less reduction in moment capacity is found for CFST members. This is due to the favorable stress distribution in CFST in bending.

梁柱的节点示意图如图 11.10(c)所示。很明显钢管混凝土构件的极限抗弯承载能力无明显降低。主要是由于钢管混凝土弯曲时产生了有利的应力分布。

(a) Unfilled RHS — local (single inward folding) failure mechanism with cracking

中空的矩形钢管——局部开裂破坏(单向内折叠)

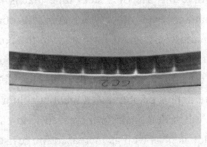

(b) RHS filled with low strength concrete — localized (single outward folding) mechanism without cracking

填充低强度混凝土的矩形钢管——局部未开裂(单向外折叠)

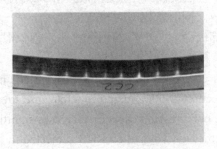

(c) RHS filled with low normal concrete — uniform (multiple outward folding) mechanism without cracking

填充低标准强度混凝土的矩形钢管——均匀无裂缝(多向外折叠)

Figure 11.10  Comparison of Different RHS

图 11.10  不同矩形钢管混凝土的对比

# New Words and Expressions　生词和短语

1. Composite Structure　　组合结构；复合机构；混合结构
2. Reinforced Concrete (RC)　　钢筋混凝土
3. concrete-filled steel tube (CFST)　　钢管混凝土
4. hollow　　*n.* 洞；山谷；窟窿
　　　　　　*adj.* 空的；中空的，空腹的；凹的

5. weld　　*n.* 焊接；焊接点
6. hot-rolled process　　热轧过程
7. cold-formed process　　冷弯过程
8. circular　　*adj.* 循环的；圆形的；间接的
9. square　　*n.* 平方；广场；正方形
　　　　　　*adj.* 平方的；正方形的；直角的；正直的
10. rectangular　　*adj.* 矩形的；成直角的
11. spatial　　*adj.* 空间的；存在于空间的；受空间条件限制的
12. spacious　　*adj.* 宽敞的，广阔的；无边无际的
13. basement　　*n.* 地下室；地窖
14. bolt　　*n.* 螺栓
15. high-rise building　　高层建筑物
16. formwork　　*n.* 量规，模架；样板
17. cantilever launching method　　悬臂发射方法
18. arch rib　　拱肋
19. tubular　　*adj.* 管状的
20. torsion　　*n.* 扭转，扭曲；转矩，[力] 扭力
21. coefficient　　*n.* [数] 系数；率；协同因素
　　　　　　　　*adj.* 合作的；共同作用的
22. ductility　　*n.* 延展性；柔软性
23. rotation angle　　旋转角；旋转角度；回转角度

# Exercises　练习

Ⅰ. Write a T in front of a statement if it is true according to the text and write an F if it is false.

1. CFST is very suitable for supporting columns in subways.

2. The hollow steel tubes can be fabricated by welding steel plates together or by hot-rolled process, or by cold-formed process.

3. The steel tubes can be filled with concrete to convert the system into a composite structure and capable of bearing the service load.

4. For concrete-filled RHS beams, neither localized outward folding nor uniform outward folding mechanism is formed without cracking.

5. The fire resistance of unprotected RHS or CHS columns is normally found to be more than 30 minutes.

Ⅱ. Complete the following sentences.

1. Please list some typical CFST section shapes commonly in practice :_____.

2. In Figure 11.1, _____ is the overall depth of the rectangular section and _____ is the steel wall thickness.

3. The popular methods being used in the CFST include _____.

4. It is well known that tubular sections have many advantages over conventional open sections, such as _____.

5. Compared with unfilled tubes, concrete-filled tubes demonstrate some properties : _____.

Ⅲ. Translate the following sentences into Chinese.

1. In recent years, such composite columns are more and more popular in high-rise or super-high-rise buildings and bridge structures.

2. The steel tube can also support a considerable amount of construction loads prior to the pumping of wet concrete, which results in quick and efficient construction.

3. It is clear that less reduction in moment capacity is found for CFST members.

4. It is well known that tubular sections have many advantages over conventional open sections.

5. Since the weight of the hollow steel tubes is comparatively small, relatively simple construction technology can be used for the erection.

Ⅳ. Translate the following sentences into English.

1. 本章内容涉及另一种类型的混凝土–钢组合结构，即钢管混凝土。

2. 近几十年来钢管混凝土的使用有增加的趋势。

3. 钢管混凝土柱具有很高的承载能力，从而可以用于宽敞的建筑。

4. 近年来，方形和矩形截面的钢管混凝土柱也在高层建筑中越来越受欢迎。

5. 与空钢管相比，钢管混凝土的承载能力、延性、在地震中的耗能能力以及抗火能力明显增强。

# Answers  答案

Ⅰ.

1. T  2. T  3. T  4. F  5. F

Ⅱ.

1. circular, square and rectangular

2. $D, t$

3. cantilever launching methods, and either horizontal or vertical "swing" methods

4. excellent strength properties, lower drag coefficients, less painting area, aesthetic merits and potential of void filling

5. increased load-carrying capacity, ductility, energy absorption during earthquakes and

increased fire resistance

Ⅲ.

1. 近年来，这样的组合柱在高层或超高层建筑和桥梁结构中越来越流行。
2. 在倒入湿混凝土之前钢管也可以支持相当数量的施工荷载，从而快速、高效地施工。
3. 很显然，在钢管混凝土构件中发现可以减少极限弯矩。
4. 众所周知，管状部分和传统的开口部分相比有许多优点。
5. 由于空心钢管重量比较小，相对简单的建筑技术可用于施工。

Ⅳ.

1. This chapter deals with another type of concrete–steel composite construction, namely concrete-filled steel tubes.
2. There is an increasing trend in using concrete-filled steel tubes in recent decades.
3. CFST columns have very high load-bearing capacity, which thus can be used in spacious construction.
4. In recent years, CFST columns with square and rectangular sections are also becoming popular in high-rise buildings.
5. Compared with unfilled tubes, concrete-filled tubes demonstrate increased load-carrying capacity, ductility, and energy absorption during earthquakes as well as increased fire resistance.

# Chapter 12  Pavement Design
# 第 12 章  道 路 设 计

## 12.1  Introduction  引言

Hard surfaced pavements, which make up about 60 percent of U.S. roads and 70 percent of Washington State roads are typically categorized into flexible and rigid pavements:

占美国 60%和占华盛顿州 70%的道路都采用了硬质路面，这些硬质路面又可以分为柔性路面和刚性路面：

Flexible pavements are those which are surfaced with bituminous (or asphalt) materials (see Figure 12.1). These types of pavements are called "flexible" since the total pavement structure "bends" or "deflects" due to traffic loads. A flexible pavement structure is generally composed of several layers of materials which can accommodate this "flexing".

柔性路面指的是表面为沥材料(或沥青质类材料)的路面(见图 12.1)。由于在交通荷载作用下，整体道路会出现"弯曲"或"弯沉"，所以这类道路被称为"柔性路面"。柔性路面一般由几层柔性良好的材料组成。

Rigid pavements are those which are surfaced with portland cement concrete (PCC) (see Figure 12.2). These types of pavements are called "rigid" because they are substantially stiffer than flexible pavements due to PCC's high stiffness.

刚性路面表层为普通水泥混凝土(PCC)的道路(见图 12.2)。由于普通水泥混凝土的刚度大，比柔性道路更坚硬，因此这类路面被称为"刚性路面"。

Figure 12.1  Flexible Pavement
图 12.1  柔性路面

Figure 12.2  Rigid Pavement
图 12.2  刚性路面

Each of these pavement types distributes load over the subgrade in a different fashion. Rigid pavement, because of PCC's high stiffness, tends to distribute the load over a relatively wide area of subgrade (see Figure 12.3). The concrete slab itself supplies most of a rigid pavement's structural capacity. Flexible pavement uses more flexible surface course and distributes loads over a smaller area. It relies on a combination of layers for transmitting load to the subgrade (see Figure 12.4).

这两种路面类型通过不同的方式把荷载传到路基上。刚性路面由于其刚度比较大，所以，它更趋向于把荷载分布到一个更大的区域传到地基上(见图 12.3)。而水泥混凝土则提供了大部分的结构承载力。柔性路面通过更有弹性的面层把荷载分布到一个较小的区域，它主要依靠路面结构各个层次的组合把荷载传递到路基上(见图 12.4)。

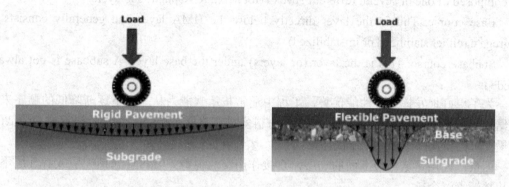

Figure 12.3   Rigid Pavement Load Distribution   Figure 12.4   Flexible Pavement Load Distribution
图 12.3   刚性路面荷载传递                图 12.4   柔性路面荷载传递

In general, both flexible and rigid pavements can be designed for long life (e.g., in excess of 30 years) with only minimal maintenance. Both types have been used for just about every classification of road. Certainly there are many different reasons for choosing one type of pavement or the other, some practical, some economical, and some political. As a point of fact, 93 percent of U.S. paved roads and about 87 percent of Washington State paved roads are surfaced with bituminous (asphalt) materials.

一般来说，不管是刚性路面还是柔性路面，在最低维修的情况下，都可以达到其设计使用年限(超过 30 年)。这两种路面类型都已经被用到各种道路工程中，当然，人们在选择使用某种道路类型时，考虑更多的是这条道路的实用性、经济性和政治性。而实际上，美国 93%的公路和华盛顿 87%的公路都是选用的沥青(或沥青类材料)路面。

## 12.2  Flexible Pavement  柔性路面

Flexible pavements are so named because the total pavement structure deflects, or flexes, under loading. A flexible pavement structure is typically composed of several layers of materials. Each layer receives the loads from the above layer, spreads them out, and then passes on these loads to the next layer below. Thus, the further down in the pavement structure a particular layer is, the less load (in terms of force per area) it must carry.

因为在荷载作用下，路面会发生弯曲或弯沉变形，柔性路面的名称由此而来。一个典型的柔性路面结构是由几层不同的路面材料组成的。路面结构中的每一层都受到上面一层传来的荷载，然后将其分散，并传递给其下一层。因此，在路面结构中，就一个区域而言，越深的部位所承受的荷载就越少。

In order to take maximum advantage of this property, material layers are usually arranged in order of descending load bearing capacity with the highest load bearing capacity material (and most expensive) on the top and the lowest load bearing capacity material (and least expensive) on the bottom. The typical flexible pavement structure consists of:

Surface course. This is the top layer and the layer that comes in contact with traffic. It may be composed of one or several different HMA (Hot Mixture Asphalt) sublayers.

Base course. This is the layer directly below the HMA layer and generally consists of aggregate (either stabilized or unstabilized).

Subbase course. This is the layer (or layers) under the base layer. A subbase is not always needed.

为了更好地发挥这种性能优势，工程中通常按承载能力从上到下逐渐降低的顺序来布置结构层，把承载能力最高(价格最贵)的材料安排在路面结构的上面，而把承载力最低(价格最便宜的)的材料安排在路面结构的下面。柔性路面结构包括：

面层：面层是道路最上面的一层，直接作用交通荷载。它可能包括一层或者几层不同的热拌沥青混合料。

基层：基层位于热拌沥青混合料面层以下，它通常由各种骨料(包括稳定骨料和不稳定骨料)组成。

垫层：垫层位于路面基层以下。道路中有时不设置垫层。

There are many different types of flexible pavements. This section covers three of the more common types of HMA mix types used in the U.S. HMA mix types differ from each other mainly in maximum aggregate size, aggregate gradation and asphalt binder content/type. This chapter focuses on dense-graded HMA in most flexible pavement sections because it is the most common HMA pavement material in the U.S. This section provides a brief exposure to:

在工程中有许多不同类型的柔性面层。在美国，有三种很常见的热拌沥青混合料混合类型。这三种类型，是由骨料最大粒径、骨料级配和沥青混合料类型的不同而划分的。本章主要介绍了在柔性路面中常用的密级配热拌沥青混合料，因为这是美国最常见的热拌沥青混合料路面材料。这一部分提供了一个简短陈述：

Dense-graded HMA. Flexible pavement information in this chapter is generally concerned with dense-graded HMA. Dense-graded HMA is a versatile, all-around mix making it the most common and well-understood mix type in the U.S.

密级配热拌沥青混合料：本章中的柔性路面数据基本上都与密集配热拌沥青混合料有关。密集配热拌沥青混合料是一种通用的综合的混合料，它的全面混合工艺使其成为美国最受欢迎的热拌沥青混合料类型。

Stone mastic asphalt (SMA). SMA, although relatively new in the U.S., has been used in Europe as a surface course for years to support heavy traffic loads and resist studded tire wear.

沥青玛琋脂碎石混合料(简称 SMA)：沥青玛琋脂碎石混合料在美国相对来说出现得较晚，但是，它在欧洲却使用了好多年了，通常被用来承载大型交通荷载并抵抗轮胎磨损带来的变形。

Open-graded HMA. This includes both open-graded friction course (OGFC) and asphalt treated permeable materials (ATPM). Open-graded mixes are typically used as wearing courses (OGFC) or underlying drainage layers (ATPM) because of the special advantages offered by their porosity.

开级配热拌沥青混合料：开级配热拌沥青混合料包括排水式开级配磨耗层(简称 OGFC)和排水式沥青稳定碎石混合料(简称 ATPM)。开级配沥青混合料由于孔隙率的特殊性质而被用在磨耗层或者是排水层。

### 1. Dense-Graded Mixes 密级配骨料混合物

A dense-graded mix (see Figure 12.5) is a well-graded HMA mixture intended for general use. When properly designed and constructed, a dense-graded mix is relatively impermeable. Dense-graded mixes are generally referred to by their nominal maximum aggregate size. They can further be classified as either fine-graded or coarse-graded. Compared with coarse-graded, there will be more excellent and like sand aggregate.

通常使用的密级配混合料(见图 12.5)是一种级配良好的热拌沥青混合料。通过合理的设计和施工，它的抗渗性很强。密级配混合料更多的是通过其公称最大骨料粒径判别，它们可以细分为细粒式和粗粒式。和粗粒式相比细粒式有更多优良的和像沙子一样的骨料。

Figure 12.5  Pavement Adopting Dense-graded Mix
图 12.5  应用密级配混合物的道路

Purpose: Dense-graded mixes are suitable for all pavement layers and for all traffic conditions. They work well for structural, friction, leveling and patching needs.

Materials: Well-graded aggregate, asphalt binder (with or without modifiers), RAP.

Mix Design: Superpave, Marshall or Hveem procedures.

应用：密级配混合料适用于所有的路面层和各种交通条件，在路面结构道路磨损、道路平整和道路修补上都有很好的使用效果。

材料：级配良好的骨料料、沥青混合料(含有或未含有改性剂)、再生沥青路面材料(RAP)。

配合比设计：高性能沥青路面设计，马歇尔及海姆配合比设计。

## 2. Stone Mastic Asphalt (SMA) Mixes　沥青玛琋脂碎石混合料

Stone mastic asphalt (SMA) is a gap-graded HMA that is designed to maximize deformation (rutting) resistance and durability by using a structural basis of stone-on-stone contact (see Figure 12.6). Because the aggregates are all in contact, rut resistance relies on aggregate properties rather than asphalt binder properties.

沥青玛琋脂碎石混合料(简称 SMA)是一种间断级配的热拌沥青混合料，它设计的目的在于抵抗最大变形(车辙)能力，并通过使用骨料咬合结构以提高耐久性(见图 12.6)。由于骨料都是互相咬合的，抵抗变形的能力取决于骨料的性能而不是沥青黏合剂的性能。

Figure 12.6　Pavement Adopting SMA
图 12.6　沥青玛琋脂碎石混合料路面

Since aggregates do not deform as much as asphalt binder under load, this stone-on-stone contact greatly reduces rutting. SMA is generally more expensive than a typical dense-graded HMA (about 20% ~ 25%) because it requires more durable aggregates, higher asphalt content and, typically, a modified asphalt binder and fibers. In the right situations it should be cost-effective because of its increased rut resistance and improved durability. SMA, originally developed in Europe to resist rutting and studded tire wear, has been used in the U.S. since about 1990.

因为在承受荷载时，骨料的变形并没有沥青那么大，所以这种骨料咬合结构会极大地减少车辙。沥青玛琋脂碎石混合料一般比标准密级配热拌沥青混合料造价高(高出20%~25%)，因为它需要更为优质的骨料、更高的沥青含量，通常还必须有改进的沥青黏合剂及纤维。但在正常情况下沥青玛琋脂碎石混合料是经济的，因为它有更强的抵抗车辙变形的能力，并且提高了耐久性。该材料最初在欧洲开发，用于抵抗车辙的形成和增大车轮摩擦，在 1990 年前后应用于美国道路。

Purpose: Improved rut resistance and durability. Therefore, SMA is almost exclusively used for surface courses on high volume interstates and U.S. roads.

Materials: Gap-graded aggregate (usually from coarse aggregate, manufactured sands and mineral filler all combined into a final gradation), asphalt binder (typically with a modifier).

用途：沥青玛琋脂碎石混合料用来提高道路抵抗车辙的能力和耐久性，因此在大多数美国国道和州际公路中，该材料几乎只用于道路面层。

材料：间断级配的骨料(通常用粗骨料、砂及矿物质填料共同合成级配)、沥青结合料(通常加入改性剂)。

Mix Design: Superpave or Marshall procedures with modifications.

配合比设计：高性能沥青路面或者改进的马歇尔法。

Other Information: Because SMA mixes have a high asphalt binder content (on the order of 6 percent), as the mix sits in the HMA storage silos, transport trucks, and after it is placed, the asphalt binder has a tendency to drain off the aggregate and down to the bottom — a phenomenon known as "mix draindown". Mix draindown is usually combated by adding cellulose or mineral fibers to keep the asphalt binder in place. Cellulose fibers are typically shredded newspapers and magazines, while mineral fibers are spun from molten rock. A laboratory test is run during mix design to ensure the mix is not subject to excessive draindown. In mix design a test for voids in the coarse aggregate is used to ensure there is stone-on-stone contact. Other reported SMA benefits include wet weather friction (due to a coarser surface texture), lower tire noise (due to a coarser surface texture) and less severe reflective cracking. Mineral fillers and additives are usually added to minimize asphalt binder drain-down during construction, increase the amount of asphalt binder used in the mix and to improve mix durability.

其他说明：因为沥青玛琋脂碎石混合料的沥青黏结剂用量大(近 6%)，在它被置于热拌沥青混合料存储筒仓、运料卡车以及在混合料放置以后，沥青黏结剂会从骨料当中流失到底部——这种现象叫作"垂流现象"。对混合料垂流现象的控制一般是通过加入化学纤维或矿物纤维来防止沥青混合料流走。纤维通常是切碎的回收报纸和杂志，或者采用岩石加工成的矿物纤维。实验室试验要在配合比设计的过程中进行，以使其不受过度垂流的影响。在配合比设计中，对粗骨料孔隙率的检测是用来确认骨架结构的形成。沥青玛琋脂碎石混合料的其他特点还有：增大的湿滑路面摩擦力(加强表面纹理的粗糙度)、降低轮胎噪声以及减小反射裂缝。在施工过程中，一般通过加入矿物质掺合料和添加剂来减少沥青的垂流现象，通过增加沥青黏结剂的比重来提高混合料的耐久性。

### 3. Open-graded Mixes　开级配混合料

An open-graded HMA mixture is designed to be water permeable (dense-graded and SMA mixes usually are not permeable). Open-graded mixes use only crushed stone (or gravel) and a small percentage of manufactured sands. There are three types of open-graded mixes typically used in the U.S.:

开级配热拌沥青混合料是透水的(密级配混合料和沥青玛琋脂碎石混合料通常是不透水的)。开级配混合料只用碎石(或沙砾)和比重很小的砂。以下为三种在美国使用的具有代表性的开级配混合物：

(1) Open-graded friction course (OGFC). Typically 15 percent air voids, no minimum air voids specified, lower aggregate standards than PEM.

(2) Porous European mixes (PEM). Typically 18～22 percent air voids, specified minimum

air voids, higher aggregate standards than OGFC and requires the use of asphalt binder modifiers.

(3) Asphalt treated permeable bases (ATPB). Less stringent specifications than OGFC or PEM since it is used only under dense-graded HMA, SMA or PCC for drainage.

(1) 排水式开级配磨耗层(OGFC)：孔隙比一般为 15%，没有限定最小孔隙，骨料标准低于欧洲多孔混合料。

(2) 欧洲多孔混合料(PEM)：孔隙比一般为 18%～22%，指定最小孔隙高于 OGFC 并且需要加入沥青改良剂。

(3) 排水式沥青稳定碎石混合料(ATPB)：比 OGFC 或 PEM 的要求更宽松，因为它只用于密级配热拌沥青混合料、沥青玛琋脂碎石混合料或 PCC 的下层的排水。

Purpose: OGFC and PEM — Used as for surface courses only. They reduce tire splash/spray in wet weather and typically result in smoother surfaces than dense-graded HMA. Their high air voids trap road noise and thus reduce tire-road noise by up to 50 percent (10 dB). ATPB — Used as a drainage layer below dense-graded HMA, SMA or PCC.

用途：OGFC 和 PEM 只用于表层。两者可以减小在降水天气的轮胎溅水，并且比密级配热拌沥青混合料路面更平整。这两种混合料较高的孔隙比可以控制道路上的噪声并将轮胎和道路的噪声减小 50%(10 分贝)。ATPB——用于密级配热拌沥青混合料、沥青玛琋脂碎石混合料或普通水泥混凝土的下层排水层。

Materials: Aggregate (crushed stone or gravel and manufactured sands), asphalt binder (with modifiers). Mix Design: Less structured than for dense-graded or SMA mixes. Open-graded mix design generally consists of ①material selection; ②gradation; ③compaction and void determination and ④asphalt binder drain-down evaluation.

材料：骨料(碎石或者沙砾、砂)、沥青黏结剂(加改良剂)。配合比设计：相较密级配或者沥青玛琋脂碎石混合料结构化程度低。开级配混合物设计一般由以下步骤组成：①原材料选择；②骨料级配；③压实度和孔隙比测定；④测定沥青黏结剂的垂流程度。

Other Information: Both OGFC and PEM are more expensive per ton than dense-graded HMA, but the unit weight of the mix when in-place is lower, which partially offsets the higher per-ton cost. The open gradation creates pores in the mix, which are essential to the mix's proper function. Therefore anything that tends to clog these pores, such as low-speed traffic, excessive dirt on the roadway or deicing sand, should be avoided.

其他说明：OGFC 和 PEM 的每吨造价均比密级配热拌沥青混合料高，但在完工后两者混合的单位重量更轻，从而抵消了一部分费用。开级配会在混合料中形成一些小孔，这些小孔对混合料功能的发挥起着至关重要的作用。因此，应该尽量避免孔隙可能被阻塞的各种因素，如车辆的低速行驶、道路上过多的沙土和尘土。

## 12.3　Rigid Pavement　刚性路面

Rigid pavements are so named because the pavement structure deflects very little under loading due to the high modulus of elasticity of their surface course. A rigid pavement structure is typically composed of a PCC surface course built on top of either the subgrade or an underlying base course. Because of its relative rigidity, the pavement structure distributes loads over a wide area with only one, or at most two, structural layers.

The typical rigid pavement structure consists of:

Surface course. This is the top layer, which consists of the PCC slab.

Base course. This is the layer directly below the PCC layer and generally consists of aggregate or stabilized subgrade.

Subbase course. This is the layer (or layers) under the base layer. A subbase is not always needed and therefore may often be omitted.

面层在荷载的作用下由于具有很高的弹性模量使得路面结构的变形非常小，故这种道路称为刚性路面。刚性路面结构通常由普通水泥混凝土面层、基层和垫层组成。因为其相对刚性，这种路面结构仅通过一个或两个结构层中的一个宽大的区域抵抗荷载。

标准的刚性路面由以下几部分组成：

面层：这是顶端的一层，由一层普通水泥混凝土材料构成。

基层：在普通水泥混凝土层的下方，一般由骨料或稳定的路基构成。

垫层：此层为路基层下方的一层或几层。有时也可以不设置垫层。

Almost all rigid pavement is made with PCC, thus this chapter only discusses PCC pavement. Rigid pavements are differentiated into three major categories by their means of crack control:

Jointed plain concrete pavement (JPCP). This is the most common type of rigid pavement. JPCP controls cracks by dividing the pavement up into individual slabs separated by contraction joints. Slabs are typically one lane wide and between 3.7m and 6.1m long. JPCP does not use any reinforcing steel but does use dowel bars and tie bars.

Jointed reinforced concrete pavement (JRCP). As with JPCP, JRCP controls cracks by dividing the pavement up into individual slabs separated by contraction joints. However, these slabs are much longer (as long as 15m) than JPCP slabs, so JRCP uses reinforcing steel within each slab to control within-slab cracking. This pavement type is no longer constructed in the U.S. due to some long-term performance problems.

Continuously reinforced concrete pavement (CRCP). This type of rigid pavement uses reinforcing steel rather than contraction joints for crack control. Cracks typically appear ever 1.1~2.4m; and are held tightly together by the underlying reinforcing steel.

几乎所有的道路都是用普通水泥混凝土制成的，因此本章只讨论普通水泥混凝土路面。刚性道路按照裂缝控制的方法分为三类：

预留接缝的无筋混凝土路面(JPCP)。这是刚性路面最为常见的种类。JPCP 通过预留伸缩缝将路面分成独立的板块来控制裂缝。每个板块通常为一个车道宽，长度在 3.7~6.1m 之间。JPCP 不使用钢筋，用的是插筋和系杆。

钢筋混凝土路面(JRCP)。正如预留接缝的无筋混凝土路面那样，JRCP 也通过预留伸缩缝将路面分成独立的板块来控制裂缝。然而，这种路面的板块比 JPCP 的板块更长(15m)，因此 JRCP 在每一块板内使用钢筋来控制板内裂缝。这种类型的道路由于一些耐久性问题已不在美国修建。

连续配筋混凝土路面(CRCP)。这种刚性道路用钢筋而非伸缩缝来控制裂缝。出现的裂缝通常在 1.1~2.4m 之间，并且被下面的钢筋牢牢地约束在一起。

## 1. Jointed Plain Concrete Pavement (JPCP)　预留接缝的无筋混凝土路面(JPCP)

Jointed plain concrete pavement (JPCP) uses contraction joints to control cracking and does not use any reinforcing steel. Transverse joint spacing is selected so that temperature and moisture stresses do not produce intermediate cracking between joints. This typically results in a spacing no longer than about 6.1m. Dowel bars are typically used at transverse joints to assist in load transfer. Tie bars are typically used at longitudinal joints.

预留接缝的无筋混凝土路面(JPCP)通过伸缩缝来控制裂缝，并且不使用任何钢筋。通过设置横向伸缩缝使其在温度和湿度的影响下不会在每一区段之间产生裂缝。这就通常使间距短于6.1m。插筋通常用在伸缩缝处来辅助荷载的传递。系杆通常用于纵向接头。

Crack Control: Contraction joints, both transverse and longitudinal.

Joint Spacing: Typically between 3.7m and 6.1m. Due to the nature of concrete, slabs longer than about 6.1m will usually crack in the middle. Depending upon environment and materials slabs shorter than this may also crack in the middle.

Reinforcing Steel: None.

Load Transfer: Aggregate interlock and dowel bars. For low-volume roads aggregate interlock is often adequate. However, high-volume roads generally require dowel bars in each transverse joint to prevent excessive faulting.

Other Information: A majority of U.S. State DOTs build JPCP because of its simplicity and proven performance.

裂缝控制：横向和纵向的伸缩缝。

接缝间距：通常在 3.7~6.1m 之间。由于混凝土的材料属性，长度超过 6.1m 的路面通常会在中间产生裂缝。长度短于 6.1m 的路面也可能由于环境和材料因素而在中间产生裂缝。

钢筋：不应用钢筋。

荷载传递：骨料咬合与插筋。对于低交通量的道路，依靠骨料的咬合足以保证荷载传递。而一般情况下，高交通量的道路通常在每个横向接缝处都需要设置插筋，以防止过度的弯沉破坏。

其他说明：鉴于其工艺简便性与可靠性，大多数美国的州级交通运输部都选择建造JPCP路面的道路。

## 2. Jointed Reinforced Concrete Pavement (JRCP)　钢筋混凝土路面(JPCP)

Jointed reinforced concrete pavement (JRCP) uses contraction joints and reinforcing steel to control cracking. Transverse joint spacing is longer than that for JPCP and typically ranges from about 7.6 m to 15.2 m. Temperature and moisture stresses are expected to cause cracking between joints, hence reinforcing steel or a steel mesh is used to hold these cracks tightly together. Dowel bars are typically used at transverse joints to assist in load transfer while the reinforcing steel/wire mesh assists in load transfer across cracks.

钢筋混凝土路面(JPCP)通过伸缩缝和钢筋来控制裂缝。横向接缝间距要比JPCP的长，在 7.6~15.2m 之间。温度和湿度会导致连接处产生裂缝，因此要用钢筋和钢筋网来把这些裂缝牢牢约束住。插筋通常用在横缝处来辅助荷载传递，同时，钢筋或钢丝网辅助伸缩缝间的荷载传递。

Crack Control: Contraction joints as well as reinforcing steel.

裂缝控制：伸缩缝和钢筋共同作用。

Joint Spacing: Longer than JPCP and up to a maximum of about 15 m. Due to the nature of concrete, the longer slabs associated with JRCP will crack.

接缝间距：比 JPCP 的要长，最大长度在 15m 左右。由于混凝土的材料属性，路面的长度越长，JRCP 路面越容易产生裂缝。

Reinforcing Steel: A minimal amount of steel is included in mid-slab to hold cracks tightly together. This can be in the form of deformed reinforcing bars or a thick wire mesh.

钢筋：板中间要用最少量的钢筋将裂缝紧紧地约束在一起。可以通过钢筋或者粗钢丝网来控制变形。

Load Transfer: Dowel bars and reinforcing steel. Dowel bars assist in load transfer across transverse joints while reinforcing steel assists in load transfer across mid-panel cracks.

荷载传递：插筋和钢筋。插筋保证伸缩缝之间的荷载传递，钢筋控制路面区块中间裂缝的荷载传递。

Other Information: During construction of the interstate system, most agencies in the Eastern and Midwestern U.S. built JRCP. Today only a handful of agencies employ this design. In general, JRCP has fallen out of favor because of inferior performance when compared to JPCP and CRCP.

其他说明：在州际系统的建立过程中，大多数美国东部和中西部的政府都选择修建钢筋混凝土路面的道路。今天只有很少的政府机关选择这种方案。一般来讲，预留接缝的无筋混凝土路面由于在性能上不及钢筋混凝土路面和连续配筋混凝土路面的道路，已经不受政府机关的青睐。

### 3. Continuously Reinforced Concrete Pavement (CRCP) 连续配筋混凝土路面(CRCP)

Continuously reinforced concrete pavement does not require any contraction joints. Transverse cracks are allowed to form but are held tightly together with continuous reinforcing steel. Research has shown that the maximum allowable design crack width is about 0.5mm to protect against spalling and water penetration. Cracks typically form at intervals of 1.1~2.4m. Reinforcing steel usually constitutes about 0.6~0.7 percent of the cross-sectional pavement area and is located near mid-depth in the slab. Typically, No. 5 and No. 6 deformed reinforcing bars are used.

连续配筋混凝土路面不需要设置任何的伸缩缝。可以产生横向裂缝，但这些横向裂缝被连续的钢筋牢牢约束住。研究表明允许设计最大裂缝宽度约为 0.5mm，以防止剥离和透水。裂缝的长度一般在 1.1~2.4m 之间。钢筋通常占道路横断面面积的 0.6%~0.7%，并且位于路面区块的中部附近。一般使用 5 号或者 6 号变形钢筋。

During the 1970's and early 1980's, CRCP design thickness was typically about 80 percent of the thickness of JPCP. However, a substantial number of these thinner pavements developed distress sooner than anticipated and as a consequence, the current trend is to make CRCP the same thickness as JPCP. The reinforcing steel is assumed to only handle nonload-related stresses and any structural contribution to resisting loads is ignored.

在 20 世纪 70 年代到 80 年代早期，CRCP 道路的设计厚度约为 JPCP 道路设计厚度的 80%。然而，大量 CRCP 的道路出现险情的时间比预料得要更早。因此，目前趋向于使 CRCP 与 JPCP 接近于等厚。增强钢筋传递非荷载应力，其对承担结构荷载不起贡献作用。

Crack Control: Reinforcing steel.

裂缝控制：钢筋。

Joint Spacing: Not applicable. No transverse contraction joints are used.

接缝间距：不适用。没有使用横向伸缩缝。

Reinforcing Steel: Typically about 0.6 ~ 0.7 percent by cross-sectional area.

钢筋：通常占横断面面积的 0.6%~0.7%。

Load Transfer: Reinforcing steel, typically No. 5 or 6 bars, grade 60.

荷载传递：钢筋，一般为 5 号或 6 号筋，60 级。

Other Information: CRCP generally costs more than JPCP or JRCP initially due to increased quantities of steel. Further, it is generally less forgiving of construction errors and provides fewer and more difficult rehabilitation options. However, CRCP may demonstrate superior long-term performance and cost-effectiveness. Some agencies choose to use CRCP designs in their heavy urban traffic corridors.

其他说明：CRCP 道路由于用钢量多于 JPCP 道路或 JRCP 道路，故造价比后两者要高。此外，这种道路在施工中的容错率更低且更难补救。然而，CRCP 道路也会展示出其长期使用的优异性能和成本效益。一些政府机关会选择将 CRCP 道路设计用于城市交通量大的道路。

# Exercises 练习

Ⅰ. Write a T infront of a statement if it is true according to the text write an F if it is false.

1. These types of pavements are called "flexible" since the total pavement structure "bends" or "deflects" due to traffic loads.

2. Subbase course. This is the layer under the base layer. A subbase is not always needed.

3. A dense-grade mix is a well-graded HMA mixture intended for general use.

4. The open gradation creates pores in the mix, which are essential to the mix's proper function.

5. Rigid pavements are differentiated into two major categories.

Ⅱ. Complete the following sentences.

1. Flexible pavements are so named because _____.

2. The typical flexible pavement structure consist of _____, _____, _____.

3. There are three types of open-graded mixes typically used in the U.S: _____, _____, _____.

4. Rigid pavements are so named because _____.

5. Joined plain concrete pavement (JPCP) is the most common type of rigid pavement, JPCP controls cracks by _____.

Ⅲ. Translate the following sentences into Chinese.

1. The flexible pavement consists of a relatively thin wearing surface built over a base course, and subbase course, and they rest upon the compacted subgrade.

2. Rigid pavement are made up of portland cement concrete and may not have a base course between the pavement and the subgrade.

3. There are many different types of flexible pavements.

4. Soil type has a greater influence on the thickness of flexible pavements than on the thickness of rigid pavements.

5. Pavements design consists of two broad categories-design of the paving mixture and structural design of the pavements components.

Ⅳ. Translate the following sentences into English.

1. 硬质路面又可以分为柔性路面和刚性路面。
2. 开级配热拌沥青混合料是透水的。
3. 插筋通常用在伸缩缝外来辅助荷载的传递。
4. 系杆通常用在纵向接头。
5. 钢筋混凝土路面(JRCP)通过伸缩缝和钢筋来控制裂缝。

# Answers  答案

Ⅰ.

1. T  2. T  3. T  4. T  5. F

Ⅱ.

1. the total pavements structure deflects, or flexes, under loading

2. surface course    Base course    subbase course

3. Open-graded friction course    Porous European mixes    Asphalt treated permeable bases

4. the pavement structure deflects very little under loading due to the high modulus of elasticity of their surface course

5. dividing the pavement up into individual slabs separated by contraction joints

III.
1. 柔性路面是由建在基层的相对薄的磨耗层、基层和底层组成,他们位于压实路基上。
2. 刚性路面是由硅酸盐水泥混凝土和路面和路基之间可有可无的基层组成。
3. 在工程中有许多不同类型的柔性面层。
4. 土的种类对柔性路面的影响比对刚性路面的更大。
5. 路面设计由铺路混合料和路面组成部分结构设计两大种类设计组成。

IV.
1. Sate roads are typically categorized into flexible and rigid pavements.
2. An open-graded HMA mixture is designed to be water permeable.
3. Dowel bar are typically used at transverse joints to assist in load transfer.
4. Tie bars are typically used at longitudinal joints.
5. Jointed reinforced concrete pavement (JRCP) uses contraction joints and reinforcing steel to control cracking.

# Chapter 13　Bridge
# 第 13 章　桥

## 13.1　Introduction　引言

A bridge is a structure providing passage over an obstacle such as a valley, road, railway, canal, river, without closing the way beneath. The required passage may be for road, railway, canal, pipeline, cycle track or pedestrians.

桥是一种工程构造物，它提供了跨越山谷、道路、铁路、沟渠、河流等障碍物的通道，且又不需要封堵障碍物。这种通道可用于道路、铁路、沟渠、管线、自行车道或人行道上。

The branch of civil engineering which deals with the design, planning construction and maintenance of bridge is known as bridge engineering. Designs of bridges vary depending on the function of the bridge and the nature of the terrain where the bridge is constructed.

桥梁工程是土木工程的分支，主要内容是桥梁结构的设计、施工组织和维修。桥梁的设计取决于桥梁的功能和建造桥梁所处的自然地形情况。

There are six main types of bridges: beam bridges, cantilever bridges, arch bridges, suspension bridges, cable-stayed bridges and truss bridges.

桥梁有六个主要的类型：梁式桥、悬臂桥、拱式桥、悬索桥、斜拉桥和桁架桥。

## 13.2　Beam Bridges　梁式桥

Beam bridges are horizontal beams supported at each end by piers, as shown in Figure 13.1. The earliest beam bridges were simple logs across streams and similar simple structures. In modern times, beam bridges are large box steel girder bridges. Weight on top of the beam pushes straight down on the piers at either end of the bridge. They are made up mostly of wood or metal.

梁式桥是指由桥墩支撑桥跨两端所形成的结构，如图 13.1 所示。最早记载的梁式桥是用于越过溪流的简单的原木和类似的简易结构。现今的梁式桥都是大箱型钢桁梁桥。桥梁上的荷载直接作用到桥跨两端的桥墩上。其材料主要为木材和金属。

Figure 13.1  Beam Bridges
图 13.1  梁式桥

The beam bridge, also known as a girder bridge, is a firm structure that is the simplest of all the bridge shapes. Both strong and economical, it is a solid structure comprised of a horizontal beam, being supported at each end by piers that endure the weight of the bridge and the vehicular traffic. Compressive and tensile forces act on a beam bridge, due to which a strong beam is essential to resist bending and twisting because of the heavy loads on the bridge. When traffic moves on a beam bridge, the load applied on the beam is transferred to the piers. The top portion of the bridge, being under compression, is shortened; while the bottom portion, being under tension, is consequently stretched and lengthened. Trusses made of steel are used to support a beam, enabling dissipation of the compressive and tensile forces. In spite of the reinforcement by trusses, length is a limitation of a beam bridge due to the heavy bridge and truss weight. The span of a beam bridge is controlled by the beam size since the additional material used in tall beams can assist in the dissipation of tension and compression.

梁式桥，也称为梁桥，是所有桥梁里形状最简单的一种。梁桥既耐用又经济，是由水平梁和支撑其两端的桥墩组成的稳固结构，用以承受桥体自重和桥面上的交通荷载。由于在较大的荷载作用下，坚固的桥梁要抵抗弯曲和扭转，所以梁桥会同时受到压力和拉力。当车辆在桥梁上移动时，荷载会从桥梁传递到桥墩上，梁桥的上部由于受压而被压缩，梁桥的下部由于受拉而被拉伸。钢桁架也可以用来作梁，抵抗压力和拉力。就算应用钢桁架，由于桥梁和桁架的自重，桥梁的跨度也一直受限。自从采用附加材料制造深梁来抵抗拉伸和压缩变形后，梁桥才突破了跨度这个限制。

Extensive research is being conducted by several private enterprises and the state agencies to improve the construction techniques and materials used for the beam bridges. The beam bridge design is oriented towards the achievement of light, strong, and long-lasting materials like reformulated concrete with high performance characteristics, fiber reinforced composite materials, electro-chemical corrosion protection systems, and more precise study of materials. Modern beam bridges use prestressed concrete beams that combine the high tensile strength of steel and the superior compression properties of concrete, thus creating a strong and durable beam bridge. Box girders are being used that are better designed to undertake twisting forces, and can make the spans longer, which is otherwise a limitation of beam bridges. The modern technique of the finite element analysis is used to obtain a better beam bridge design, with a meticulous analysis of the stress distribution, and the twisting and bending forces that may cause failure.

一些国家相关机构和企业都进行过大量的研究，来改良梁式桥的施工技术和材料。梁式桥的设计目的是达到结构的轻质高强，包括应用新型高性能混凝土、纤维混凝土、电子化学防腐系统等来实现材料的耐久。现代梁式桥使用的是预应力混凝土，这种结构应用了具有高抗拉强度的钢筋与具有优良抗压性能的混凝土，以实现桥梁的坚固和耐久。在以往为了突破结构限制，梁式桥通常设计使用箱型梁，从而可以更好地抗扭和加大桥的跨度。现在则运用有限元分析技术来优化桥梁设计，该技术具有精密的应力分布分析以及桥梁的扭转和弯曲破坏分析。

## 13.3  Cantilever Bridges  悬臂桥

A cantilever bridge is a bridge built using cantilevers, structures that project horizontally into space, supported on only one end, as shown in Figure 13.2. For small footbridges, the cantilevers may be simple beams; however, large cantilever bridges designed to handle road or rail traffic use trusses built from structural steel, or box girders built from prestressed concrete. The steel truss cantilever bridge was a major engineering breakthrough when first put into practice, as it can span distances of over 460m, and can be more easily constructed at difficult crossings by virtue of using little or no falsework.

悬臂桥是采用一端结构水平悬空、另一端仅固定在支座上的悬臂梁而组成的结构形式，如图 13.2 所示。对于小型的人行天桥来说，悬臂可能是一根简单横梁；然而应用于公路或铁路的大型悬臂桥，其结构形式则采用钢桁架或是预应力混凝土箱梁。应用钢桁架悬臂桥是桥梁工程上的一项突破性进展，它使桥梁可以跨越超过 460m 的距离，而且桥梁可以更容易在复杂的十字交叉路口修建，其优点是少量使用或者不使用脚手架。

Figure 13.2  Cantilever Bridge
图 13.2  悬臂桥

A simple cantilever span is formed by two cantilever arms extending from opposite sides of the obstacle to be crossed, meeting at the center. In a common variant, the suspended span, the cantilever arms do not meet in the center; instead, they support a central truss bridge which rests on the ends of the cantilever arms. The suspended span may be built off-site and lifted into place, or constructed in place using special traveling supports.

一个简单的悬臂桥梁由两根悬臂梁构成，这两根悬臂梁安装在障碍物的两端，跨越障碍物并在中心区连接。在共同的变动处，即悬臂跨中处，悬臂梁不能连接，通常应用中心桁架连接悬臂的末端。悬臂梁可以在加工厂生产并由吊车将其吊装到位，或者使用特殊的搬运装置将其安装到位。

A common way to construct steel truss and prestressed concrete cantilever spans is to counterbalance each cantilever arm with another cantilever arm projecting the opposite direction, forming a balanced cantilever; when they attach to a solid foundation, the counterbalancing arms are called anchor arms. Thus, in a bridge built on two foundation piers, there are four cantilever arms: two which span the obstacle, and two anchor arms which extend away from the obstacle. Because of the need for more strength at the balanced cantilever's supports, the bridge superstructure often takes the form of towers above the foundation piers. The Commodore Barry Bridge is an example of this type of cantilever bridge.

修筑钢桁架悬臂跨和预应力混凝土悬臂跨的常用方法是使每一根悬臂梁与另外一根相反方向的悬臂梁平衡，构成一个平衡的悬臂；当它们固定在坚固的基础中时，平衡的悬臂就称为锚固臂。因此，修建在两桥墩之间的桥有四种悬臂：其中两个跨越障碍物，另外两个向着远离障碍物的方向延伸。由于在平衡悬臂的支撑处需要更多的强度保证，桥的上部结构通常建造成在桥墩之上的塔状形式。巴里大桥就是这类悬臂大桥的实例。

Steel truss cantilevers support loads by tension of the upper members and compression of the lower ones. Commonly, the structure distributes the tension via the anchor arms to the outermost supports, while the compression is carried to the foundations beneath the central towers. Many truss cantilever bridges use pinned joints and are therefore statically determinate with no members carrying mixed loads.

钢桁架悬臂梁由上部的抗拉构件和下部的抗压构件来共同承担荷载。通常，拉力通过锚索传递到桁架最外侧的支撑，而压力通过中央塔传递到地基。许多桁架悬臂桥使用铰接节点，因成组成的静定桁架杆件都只承受轴力。

Prestressed concrete balanced cantilever bridges are often built using segmental construction. Some steel arch bridges are built using pure cantilever spans from each side, with neither falsework below nor temporary supporting towers and cables above. These are then joined with a pin, usually after forcing the union point apart, and when jacks are removed and the bridge decking is added the bridge becomes a truss arch bridge. Such unsupported construction is only possible where appropriate rock is available to support the tension in the upper chord of the span during construction, usually limiting this method to the spanning of narrow canyons.

预应力混凝土平衡悬臂桥通常应用分布施工来建造。这种工艺中，双向的纯悬臂跨使用的通常是钢拱，索塔与悬索之间既没有脚手架也没有临时支撑。当桁架桥桥面铺装完成后，用千斤顶将桥跨推动到指定位置，然后通过螺栓连接固定。在施工中，此种没有支撑的结构仅适用于有能够支撑抗拉上弦的岩石的地方，在跨越狭窄的峡谷时，此工艺就受限制了。

## 13.4　Arch Bridges　拱桥

An arch bridge is a bridge with abutments at each end shaped as a curved arch. Arch bridges work by transferring the weight of the bridge and its loads partially into a horizontal thrust restrained by the abutments at either side. A viaduct may be made from a series of arches, although other more economical structures are typically used today.

拱桥是一种两侧有桥墩并且形状是拱形的桥。拱桥的受力原理是：通过两侧的桥墩将其自身重力和桥上部分荷载转化为水平推力。如今尽管其他更多较为经济的结构被采用，但高架桥通常还是由一系列的拱构成。

There are some variations of arch bridges.

以下是各类型的拱桥。

1) Corbel Arch Bridges　叠涩拱桥

The corbel arch bridge (as shown in Figure 13.3) is a masonry or stone bridge where each successively higher course cantilevers slightly more than the previous course. The steps of the masonry may be trimmed to make the arch have a rounded shape. The corbel arch does not produce thrust, or outward pressure at the bottom of the arch, and is not considered a true arch. It is more stable than a true arch because it does not have this thrust. The disadvantage is that this type of arch is not suitable for large spans.

叠涩拱桥(如图 13.3 所示)是砌块或是石头砌筑成的桥梁，每一种都比前文所述的悬臂桥结构更高。砌块阶梯通常被砌筑成拱形或圆形。叠涩拱桥既不会产生推力也不会对拱顶产生向外的压力，它并不是实际意义上的拱形。它比实际的拱桥更加坚固，因为它不产生水平推力。此类型拱桥最大的缺点是不适合大跨度的桥。

Figure 13.3　Corbel Arch Bridges
图 13.3　叠涩拱桥

2) Aqueducts and Canal Viaducts　水道-运河高架桥

In some locations it is necessary to span a wide gap at a relatively high elevation, such as when a canal or water supply must span a valley. Rather than building extremely large arches, or very tall supporting columns, a series of arched structures are built one atop another, with wider structures at the base, as shown in Figure 13.4. Roman civil engineers developed the design and constructed highly refined structures using only simple materials, equipment, and mathematics. This type is still used in canal viaducts and roadways as it has a pleasing shape, particularly when spanning water, as the reflections of the arches form a visual impression of circles or ellipses.

在一些地区，工程要求跨越一个高海拔处的较宽间隙，例如，必须跨越峡谷修建运河或供应水流。这时就不是修建一个极其大的拱桥，或是很高的支撑拱柱，而是修建一系列彼此叠加在一起的大范围拱形建筑结构，如图13.4所示。罗马的土木工程师运用简单的材料和施工设备以及初级的数学知识发展了这种设计方法，并建造了高精度的高架引水桥。这种结构形式外形美观，特别是当引水时，水中拱的倒影会形成圆或椭圆的视觉影像，因此运河高架桥或公路桥中仍然应用这种桥梁结构形式。

Figure 13.4 Aqueducts and Canal Viaducts
图 13.4 水道-运河高架桥

3) Deck Arch Bridges 上承式拱桥

This type of bridge comprises an arch where the deck is completely above the arch, as shown in Figure 13.5. The area between the arch and the deck is known as the spandrel. If the spandrel is solid, usually the case in a masonry or stone arch bridge, it is call a closed-spandrel arch bridge. If the deck is supported by a number of vertical columns rising from the arch, it is known as an open-spandrel arch bridge. The Alexander Hamilton Bridge is an example of an open-spandrel arch bridge.

上承式桥的桥面完全设置在承重的拱结构之上，如图13.5所示。拱和桥面之间的区域称为拱肩。当拱桥是由砌体或石块砌成，其拱肩被填实，则被称为封闭拱肩拱桥。如果桥面是被数根垂直的在拱上竖起的桥柱所支撑，则被称为开放式拱肩拱桥。亚历山大-哈密顿大桥就是开放式拱肩桥的实例。

Figure 13.5  Deck Arch Bridges
图 13.5  上承式拱桥

4) Through Arch Bridges  中承式拱桥

This type of bridge comprises an arch which supports the deck by means of suspension cables or tie bars, as shown in Figure 13.6. The Sydney Harbour Bridge is a through arch bridge which uses a truss type arch. These through arch bridges are in contrast to suspension bridges which use the catenary in tension to which the aforementioned cables or tie bars are attached and suspended.

中承式拱桥同样是拱承载桥面，该拱结构是由悬索或拱肋组成，如图 13.6 所示。悉尼海港大桥是使用桁架拱的拱形桥。这种贯通型拱桥类似于悬索桥，使用了如前所述的受拉悬索和吊杆。

Figure 13.6  Through Arch Bridges
图 13.6  中承式拱桥

5) Tied Arch Bridges  下承式拱桥

Also known as a bowstring arch, this type of arch bridge incorporates a tie between two opposite ends of the arch. The tie is capable of withstanding the horizontal thrust forces which would normally be exerted on the abutments of an arch bridge, as shown in Figure 13.7.

下承式拱桥也被称为弓弦式拱桥，此类型拱桥在两个相反的拱末端形成连接。这个连接通常能承受被施加在拱桥桥台的水平推力，如图 13.7 所示。

Figure 13.7　Tied Arch Bridges

图 13.7　下承式拱桥

## 13.5　Suspension Bridges　悬索桥

A suspension bridge is a type of bridge in which the deck is hung below suspension cables on vertical suspenders, as shown in Figure 13.8. This type of bridge dates from the early 19th century, while bridges without vertical suspenders have a long history in many mountainous parts of the world.

悬索桥是一种典型的由悬索固定吊杆，吊杆吊住桥面的桥梁结构，如图 13.8 所示。悬索桥最早可以追溯到 19 世纪，当时世界上许多山区都存在古老的没有吊杆的悬索桥。

Figure 13.8　Suspension Bridges

图 13.8　悬索桥

This type of bridge has cables suspended between towers, plus vertical suspender cables that carry the weight of the deck below, upon which traffic crosses. This arrangement allows the deck to be level or to arc upward for additional clearance. Like other suspension bridge types, this type often is constructed without falsework.

这种类型的桥是将悬索悬挂在索塔之间，并以垂挂的悬索承担负载交通的桥面的重量。此类桥施工中桥面既可以保持水平，也可以使桥面起拱。相比于其他悬索桥，此类悬索桥施工时通常不需要脚手架。

The suspension cables must be anchored at each end of the bridge, since any load applied to the bridge is transformed into a tension in these main cables. The main cables continue beyond the pillars to deck-level supports, and further continue to connections with anchors in the ground. The roadway is supported by vertical suspender cables or rods, called hangers. In some circumstances the towers may sit on a bluff or canyon edge where the road may proceed directly to the main span, otherwise the bridge will usually have two smaller spans, running between either pair of pillars and the highway, which may be supported by suspender cables or may use a truss bridge to make this connection. In the latter case there will be very little arc in the outboard main cables.

悬索桥的悬索都被锚固在桥面每侧的末端，因此桥承受的任何外荷载都被转化成拉力传递到这些主索上。这些主索一端越过塔架固定在桥面上，一端则锚固在地基上。路面被悬索或是吊杆所支撑，被称为吊桥。某些情况下，桥索塔可能屹立在悬崖或是峡谷边缘，这些地方的路面直通到主跨；另外桥通常会有两小跨，在两个塔架或者主干线之间延伸，有悬索支撑或者使用桁架桥使其连接。在后一种情况下除了主索外还会存在一小部分的拱。

The main forces in a suspension bridge of any type are tension in the cables and compression in the pillars. Since almost all the force on the pillars is vertically downwards and they are also stabilized by the main cables, the pillars can be made quite slender, as on the Severn Bridge, near Bristol, England. In a suspended deck bridge, cables suspended via towers hold up the road deck. The weight is transferred by the cables to the towers, which in turn transfer the weight to the ground.

任何类型悬索桥主要承受的外力都是悬索的拉力和塔架的压力。因为几乎所有传到塔架上的外力都向下传递并且由主索固定，塔架可以足够细，如英国布里斯托附近的赛文桥。在悬索桥桥面上，悬索悬挂在固定于桥面上的索塔上。其自重通过悬索传到索塔，再依次把荷载传到地基上。

Assuming a negligible weight as compared to the weight of the deck and vehicles being supported, the main cables of a suspension bridge will form a parabola. One can see the shape from the constant increase of the gradient of the cable with linear distance, this increase in gradient at each connection with the deck providing a net upward support force. Combined with the relatively simple constraints placed upon the actual deck, this makes the suspension bridge much simpler to design and analyze than a cable-stayed bridge, where the deck is in compression.

假设桥面自重和其所承受的交通荷载可以忽略不计的话，悬索桥的主悬索将形成一个抛物线。可以看出随着悬索间距增大，悬索的倾斜度持续增大，在每一个与桥面连接处所增加的倾斜度都会使悬索产生一个竖直向上的反力。桥面板与相对简单的约束连接，使得悬索桥比桥面受压力的斜拉桥设计和计算分析更加简单。

## 13.6　Cable-stayed Bridges　斜拉桥

A cable-stayed bridge is a bridge that consists of one or more columns, with cables supporting the bridge deck, as shown in Figure 13.9. There are two major classes of cable-stayed bridges: In a harp design, the cables are made nearly parallel by attaching cables to various points on the tower so that the height of attachment of each cable on the tower is similar to the distance from the tower along the roadway to its lower attachment. In a fan design, the cables all connect to or pass over the top of the towers.

斜拉桥由一根或多根柱子以及支撑桥面的斜拉索组成，如图 13.9 所示。有两种主要的斜拉桥：竖琴型斜拉桥，这种斜拉桥通过将斜拉索连接到桥塔的不同位置，使每根斜拉索平行，这样就使塔上的每一根斜拉索的高度与沿其下部的塔的距离相同。在扇形设计斜拉桥中，斜拉索全部连接或是通过桥塔的顶部。

Figure 13.9　Cable-stayed Bridges
图 13.9　斜拉桥

Compared with other bridge types, the cable-stayed is optimal for spans longer than typically seen in cantilever bridges and shorter than those typically requiring a suspension bridge. This is the range in which cantilever spans would rapidly grow heavier if they were lengthened, and in which suspension cabling does not get more economical were the span to be shortened.

同其他类型的桥相比，斜拉桥优于大跨度的悬臂桥和小跨度的悬索桥。这是由于随着跨度增大，悬臂桥自重迅速增大，而小跨度的悬索桥则非常不经济。

A multiple-tower cable-stayed bridge may appear similar to a suspension bridge, but in fact is very different in principle and in the method of construction. In the suspension bridge, a large cable is made up by "spinning" small diameter wires between two towers, and at each end to anchorages into the ground or to a massive structure. These cables form the primary load-bearing structure for the bridge deck. Before the deck is installed, the cables are under tension from only their own weight. Smaller cables or rods are then suspended from the main cable, and used to support the load of the bridge deck, which is lifted in sections and attached to the suspender cables. As this is done the tension in the cables increases, as it does with the live load of vehicles or persons crossing the bridge. The tension on the cables must be transferred to the earth by the anchorages, which are sometimes difficult to construct due to poor soil conditions.

多塔型斜拉桥可能看起来与悬索桥相似，但实际上两种桥梁无论在原理上还是在施工工艺上都大相径庭。悬索桥中，两桥塔之间的大型悬索由小直径的金属丝缠绕制成，每一根悬索末端都要锚固在地基或是大型结构之中。这些悬索形成桥面基本的承重结构。在桥面安装之前，悬索只承受其自重产生的拉力。悬挂在主悬索中的小悬索或杆件主要用于承担桥面的荷载，它们部分吊装并连接在悬挂的悬索上。随着车辆或行人这些动载穿过桥梁，这些悬索的拉力也随之增大。悬索上的拉力必须由锚固构件传递到地基上，这些锚固构件往往由于地形复杂而难以施工。

## 13.7　Truss Bridges　桁架桥

A truss bridge is a bridge composed of connected elements which may be stressed from tension, compression, or sometimes both in response to dynamic loads, as shown in Figure 13.10. Truss bridges are one of the oldest types of modern bridges. A truss bridge is economical to construct owing to its efficient use of materials.

桁架桥由杆件组成，桥梁承受拉力、压力以及由动载而产生的拉压力，如图 13.10 所示。桁架桥是现代桥梁中最早期的类型之一。由于材料的高效利用，修建桁架桥更为经济。

Figure 13.10　Truss Bridges
图 13.10　桁架桥

Truss girders, lattice girders or open web girders are efficient and economical structural systems, since the members experience essentially axial forces and hence the material is fully utilized. Members of the truss girder bridges can be classified as chord members and web members. Generally, the chord members resist overall bending moment in the form of direct tension and compression and web members carry the shear force in the form of direct tension or compression. Due to their efficiency, truss bridges are built over wide range of spans. Truss bridges compete against plate girders for shorter spans, against box girders for medium spans and cable-stayed bridges for long spans.

桁架梁、格构梁或者开放式的网格梁是既经济又有高效的结构体系，由于构件只承受轴力，因此材料得到了充分的利用。桁架梁桥的构件被分为弦构件和网构件。通常，弦构件承担由直接的拉力和压力所形成的弯矩，而网构件承担由直接的拉力和压力所形成的剪力。由于其高利用率，在大跨度工程中通常使用桁架桥。在小跨度工程中桁架桥优于板桥，对于中等跨度优越于箱型梁，对于大跨度则打败了斜拉桥。

For short and medium spans it is economical to use parallel chord trusses such as Warren truss, Pratt truss, Howe truss, etc. to minimize fabrication and erection costs. Especially for shorter spans the Warren truss is more economical as it requires less material than either the Pratt or Howe trusses. However, for longer spans, a greater depth is required at the centre and variable depth trusses are adopted for economy. In case of truss bridges that are continuous over many supports, the depth of the truss is usually larger at the supports and smaller at mid-span.

对于中小跨度，通常利用平行弦桁架型式，以减少建造和安装的费用，如华伦式桁架、普拉特桁架、豪式桁架等。尤其对于小跨度，因华伦式桁架相较于其他两种桁架需要的材料更少，因而更加经济。然而在跨度较大、跨中高度较大的情况下，非平行弦桁架因其经济而被采用。对于连续多支座的桁架桥来说，通常桁架的高度在支座处较大而在跨中处较小。

## New Words and Expressions  生词和短语

1. terrain　　*n.* 地形，地面，地域，地带
2. beam bridge　　梁桥
3. cantilever bridge　　悬臂桥
4. arch bridge　　拱桥
5. suspension bridge　　悬索桥；吊桥
6. cable　　*n.* 缆索，钢丝绳
7. stay　　*n.* 拉索，拉杆
　　　　*v.* 牵拉，支撑
8. cable-stayed bridge　　斜拉桥
9. truss bridge　　桁架桥
10. horizontal beam　　水平梁
11. pier　　*n.* (桥)墩；支柱；码头
12. log　　*n.* 原木，木材，木料
13. girder　　*n.* 主梁，大梁，纵梁；桁架
14. span　　*n.* 跨距，跨径
15. long span　　大跨度

16. flyover    n. 立交桥，高架道路
17. mortar    n. 砂浆，灰浆
18. pole    n. 柱，杆
19. column    n. 圆柱；柱形物
20. tensile strength    拉伸强度，抗拉强度
21. compressive force    压力
22. bending    n. 弯曲，弯折；挠度
23. twisting    n. 扭曲；翘曲
24. composite material    复合材料
25. electro-chemical corrosion    电化学腐蚀
26. prestressed concrete    预应力混凝土
27. box girder    箱梁
28. the finite element analysis    有限元分析
29. footbridge    n. 人行桥
30. falsework    临时支架；脚手架
31. anchor    n. & v. 锚固，锚碇，固定
32. segmental construction    预制段拼装施工，节段施工
33. jack    n. 千斤顶，起重器
34. deck    v. 作桥面
           n. 桥面
35. canyon    n. 峡谷
36. abutment    n. 桥台，拱座
37. viaduct    n. 高架(跨线)桥，栈道
38. masonry    n. 砖石，砌体，圬工
39. aqueduct    n. 渡槽
40. deck arch bridge    上承式拱桥
41. spandrel    n. 拱肩，拱腹
42. through arch bridge    下承式拱桥
43. suspension cable    悬索
44. tie bar    系杆
45. tied arch bridge    系杆拱桥
46. bowstring arch    系杆拱
47. thrust force    推力
48. suspender    n. 吊杆，悬杆(索)
49. clearance    n. 净空，间距
50. the main cable    主缆
51. pillar    n.(索塔)柱
52. rod    n. 杆，拉杆
53. hanger    n. 吊杆，悬架

54. bluff　　*n.* 悬崖，陡岸
55. the main span　　主跨
56. parabola　　*n.* 抛物线
57. stiffness　　*n.* 刚度，刚性
58. aerodynamic　　*adj.* 空气动力学的
59. vibrate　　*vi.* 振动
60. live load　　活载，动载
61. multiple-tower cable-stayed bridge　　多塔斜拉桥
62. anchorage　　*n.* 锚碇
63. truss girder　　桁架梁
64. lattice girder　　井字梁
65. web　　*n.* 腹板
66. open web girder　　空腹梁
67. bending moment　　弯矩
68. plate girder　　板梁
69. parallel chord truss　　平行弦桁架
70. erection　　*n.* 建造

# Exercises　练习

Ⅰ. Write a T in front of a statement if it is true according to the text and write an F if it is false.

1. The earliest beam bridges were constructed using stones.
2. Arch bridges work by transferring the weight of the bridge and its loads partially into a horizontal thrust restrained by the abutments at either side.
3. A cable-stayed bridge is a type of bridge in which the deck is hung below suspension cables on vertical suspenders.
4. A cable-stayed bridge can achieve longer main span than any other types of bridges.
5. For bridges with short or medium spans, it is economical to use parallel chord trusses to minimize fabrication and erection costs.

Ⅱ. Complete the following sentences.

1. The branch of civil engineering which _____, planning construction and maintenance of bridge is _____.
2. The modern technique of the finite element analysis is _____, _____, _____.
3. A simple cantilever span is formed by _____, _____.
4. Prestressed concrete balanced cantilever bridges are _____.

5. An arch bridge is a bridge with _____ .

Ⅲ. Translate the following sentences into Chinese.

1. Designs of bridges vary depending on the function of the bridge and the nature of the terrain where the bridge is constructed.

2. The span of a beam bridge is controlled by the beam size since the additional material used in tall beams can assist in the dissipation of tension and compression.

3. Box girders are being used that are better designed to undertake twisting forces, and can make the spans longer, which is otherwise a limitation of beam bridges.

4. The suspended span may be built off-site and lifted into place, or constructed in place using special traveling supports.

5. Because of the need for more strength at the balanced cantilever's supports, the bridge superstructure often takes the form of towers above the foundation piers.

Ⅳ. Translate the following sentences into English.

1. 许多桁架悬臂桥使用铰接节点，组成的静定桁架杆件都只承受轴力。
2. 叠涩拱桥既不会产生推力也不会对拱顶产生向外的压力，它并不是实际意义上的拱形。
3. 如果桥面是被数根垂直的在拱上竖起的桥柱所支撑，则被称为开放式拱肩拱桥。
4. 中承式拱桥同样是拱承载桥面，该拱结构是由悬索或拱肋组成。
5. 国内现行规范对桥梁设计提出的要求是适用、经济、安全、美观，这些要求基本上包含了人们关心的所有重要的问题。

# Answers  答案

Ⅰ

1. F    2. T    3. T    4. F    5. T

Ⅱ.

1. deals with the design     known as bridge engineering

2. used to obtain a better beam bridge design

with a meticulous analysis of the stress distribution

and the twisting and bending forces that may cause failure

3. two cantilever arms extending from opposite sides of the obstacle to be crossed meeting at the center

4. often built using segmental construction

5. abutments at each end shaped as a curved arch

Ⅲ.

1. 桥梁的设计取决于桥梁的功能和建造桥梁所处的自然地形情况。
2. 自从采用附加材料制造深梁来抵抗拉伸和压缩变形后，桥梁才突破了跨度这个限制。
3. 在以往为了突破结构限制，梁式桥通常设计使用箱型梁，从而可以更好地抗扭和加大桥的跨度。
4. 悬臂梁可以在加工厂生产并且由吊车将其吊装到位，或者使用特殊的搬运装置将其安装到位。
5. 由于在平衡悬臂的支撑处需要更多的强度保证，桥的上部结构通常建造成在桥墩之上的塔状形式。

Ⅳ.

1. Many truss cantilever bridges use pinned joints and are there fore statically determinate with no members carrying mixed loads.
2. The corbel arch does not produce thrust, or outward pressure at the bottom of the arch, and is not considered a true arch.
3. If the deck is supported by a number of vertical columns rising from the arch, it is known as an open-spandrel arch bridge.
4. This type of bridge comprised an arch which supports the deck by means of suspension cables or tie bars.
5. The requirement that the domestic current norm puts forward for bridge design is the applicable, economy, safe, esthetic. These demand basically include all important problems that people care about.

# Chapter 14　Basic Concepts in Surveying
# 第 14 章　测量的基本概念

## 14.1　Introduction　引言

### 1. The Definition of Surveying　测量的定义

Surveying is the art of determining the relative positions of points on, above or beneath the surface of the earth by means of direct or indirect measurements of distance, direction and elevation.

测量是确定地表、地上及地下测点的相对位置的一门技术，可采用直接或间接的量测方法来确定测点的距离、方向及高度。

Levelling is a branch of surveying with the object of which is:
水准测量是其中的一个分支，目的是：

(1) to find the elevations of points with respect to a given or assumed datum.
确定一点相对于给定的基准点的高程。

(2) To establish points at a given elevation or at different elevations with respect to a given or assumed datum.
在一给定的高程上建立一点，或基于给定的基准点建立不同高程的点位。

In general, surveying deals with measurements in an horizontal plane where as levelling deals with measurements in a vertical plane.
一般来讲，测量是解决一个水平面的测量值，而水准测量则是完成一个垂直平面上的测量值。

### 2. Primary Divisions of Survey　测量的基本类型

The two primary divisions of survey are Plane survey and Geodetic survey.
测量主要分为两种类型:平面测量和大地测量。

1) Plane Surveying　平面测量

It is that type of surveying in which the mean surface of the earth is considered as a plane and the spheroidal shape is neglected. All triangles formed by survey lines are considered as plane triangles. The level line is considered as straight and plumb lines are considered parallel. In other words, when small areas are to be surveyed, the curvature of the earth's surface is ignored and such a surveying is called plane surveying.

平面测量假定地球表面为均匀平面而不考虑其真实球形表面的影响。由测线所构成的所有三角形均视为平面三角形。水平线被认为是直线，而铅垂线都是相互平行的。换句话说，当进行小范围的测量时，地球表面的曲率忽略不计，此类测量称为平面测量。

2) Geodetic Survey  大地测量

It is that type of surveying in which the curved shape of the earth is taken in to account. The object of geodetic survey is to determine the precise position on the surface of the earth, of a system of widely distant points which form control stations in which surveys of less precision may be referred.

该类测量是考虑地球表面的曲线形状带来的影响。其主要目的是确定地球表面的精确位置，它可根据广泛分布的点所形成的控制系统完成，在该类测量系统内，一般精度即可满足要求。

## 14.2  Classification  测量类型

### 1. Classification  测量类型

Surveys may be classified based on the nature of the field of survey, object of survey and instruments used.

基于测量的应用领域、测量目的及所使用的工具，测量分为如下几类：

1) Topographical Survey  地形测量

They are carried out determine the position of natural features of a region such as rivers, streams, hills etc. and artificial features such as roads and canals. The purpose of such surveys is to prepare maps and such maps of are called topo-sheets.

地形测量是为了获得一个地区的自然地表特征，如河流、溪流及山丘等，以及人造地表特征，如道路和运河等。测量的目的是制作地图或地形图。

2) Hydrographic Survey  水文测量

Hydro-graphic survey is carried out to determine M.S.L. (Mean Sea Level), water spread area, depth of water bodies, velocity of flow in streams, cross-section area of flow etc.

水文测量是为了确定海平面位置、水扩散面积、水体深度、流动速度及截面流量等。

3) Astronomical Survey  天文测量

The Astronomical Survey is carried out to determine the absolute location of any point on the surface of earth. The survey consists of making observations to heavenly bodies such as stars.

天文测量是确定在地球的表面上任意点的绝对位置。这类测量包括天体测量如星位等。

4) Engineering Survey  工程测量

This type of survey is undertaken whenever sufficient data is to be collected for the purpose of planning and designing engineering works such as roads, bridges and reservoirs.

工程测量要收集到足够多的量测数据，目的是为工程结构，如道路、桥梁和水库等的规划及设计所用。

## Chapter 14 Basic Concepts in Surveying

5) Archeological Survey 考古测量

This type of survey is carried out to gather information about sites that are important from archeological considerations and for unearthing relics of antiquity.

这种类型的测量是为了收集遗址的信息，对考古和古代文物的发掘非常重要。

6) Photographic Survey 遥感测量

In this type of survey, information is collected by taking photographs from selected points using a camera.

这种类型的测量是通过选点后利用相机拍摄来收集信息。

7) Aerial Survey 航空测量

In this type of survey data about large tracks of land is collected by taking photographs from an aero-plane.

这种类型的测量是通过从航空飞机上拍照来收集大范围的陆地信息。

8) Reconnaissance Survey 勘测测量

In this type of survey, data is collected by marking physical observation and some measurements using simple survey instruments.

这种类型的测量是通过利用物理观察进行标记和使用一些简单的测量工具进行测量来收集数据信息。

### 2. Units of Measurements 测量单位

Unit is the numerical standard used to measure the qualitative dimension of a physical quantity. Four fundamental concepts of mechanics are length, time mass and force. To satisfy Newton's second law of motion (ie. $F = ma$), three of the basic units may be defined arbitrarily and referred to as base units. Fourth unit is the derived unit. The system of unit based on length, time and force is called Gravitational System and that based on length, time and mass is called as Absolute System.

单位是用来衡量物理量大小的数量标准。四个基本力学单位分别为长度、时间、质量和力。为了满足"牛顿第二运动定律"（即 $F=ma$），任意规定了三个基本单位，第四个单位是推导出来的。基于长度、时间和力的单位系统称为重力系统，基于长度、时间和质量的单位系统称为绝对系统。

International system of units (SI) is a form of absolute system of unit. In this the units of length, mass and time are meter (m), kilogram (kg) and seconds (s). The unit of force is derived unit called Newton (N). and is defined as the force which give an acceleration of 1 m/s$^2$ to a mass of 1 kg weight of a body or force of gravity exerted on that body, should like any other force be expressed in Newton. Weight of a body of mass 1 kg is 9.81 N ($W= mg$=9.81). Some of prefix used in SI units are shown in Table 14.1.

国际单位制(SI)即为一种绝对单位制。该系统中，长度、质量和时间的单位分别是米(m)、千克(kg)和秒(s)。力的单位是导出单位，称为牛顿(N)，也是对 1kg 的物体施加了 1m/s$^2$ 加速度所施加的力，或者是将类似任何其他牛顿力自重施加到该物体上。质量为 1kg 的物体的重量约为 9.81 N($W=mg$=1×9.81=9.81N)。一些用于 SI 单位的缩写如表 14.1 所示。

Table 14.1  Prefix Used in SI Units

表 14.1  国际单位制中的一些前缀

| | | |
|---|---|---|
| 1 | T (Tetra) | $1 \times 10^{12}$ |
| 2 | G (Giga) | $1 \times 10^{9}$ |
| 3 | M (Mega) | $1 \times 10^{6}$ |
| 4 | k (kilo) | $1 \times 10^{3}$ |
| 5 | h (hecto) | $1 \times 10^{2}$ |
| 6 | c (centi) | $1 \times 10^{-1}$ |
| 7 | m (mili) | $1 \times 10^{-3}$ |
| 8 | μ (micro) | $1 \times 10^{-6}$ |
| 9 | n (nano) | $1 \times 10^{-9}$ |

## 14.3  Principles of Surveying  测量原理

The fundamental principle upon which the various methods of plane surveying are based can be stated under the following two aspects.

用于平面测量的基本理论可以从下列两个方面阐述。

(1) Location of a point by measurement from two points of reference.

通过测量两个参考点来确定另一个点的位置。

According to this principle, the relative position of a point to be surveyed should be located by measurement from at least two points of reference, the positions of which have already been fixed, as shown in Figure 14.1.

根据这一原理，一个点的相对位置要通过至少两个已经固定位置的参考点来确定，如图 14.1 所示。

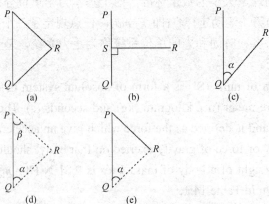

Figure 14.1  Location of A Point by Measurement from two Points of Reference

图 14.1  根据两参考点确定另一点的位置

## Basic Concepts in Surveying

If $P$ and $Q$ are the two reference points on the ground, any other point, such as $R$, can be located by any of the direct methods shown in the above figures. But, although a single method is sufficient to locate the relative position of R with respect to reference points $P$ and $Q$, it is necessary to adopt at least any two methods to fix the position of point $R$. While the measurements made in the either of the first method or second method will be helpful in locating the point $R$, the measurements made in the other method will act as a check.

如果 $P$ 和 $Q$ 是在地面上的两个参考点，其他任何点，如 $R$，可以通过上图任意一个直接的方法来确定其位置。但是，尽管一种方法足以确定 $R$ 的相对于参考点 $P$ 和 $Q$ 的位置，但仍有必要采用至少两种方法来确定 $R$ 点的位置。无论用第一或第二种方法都有助于确定 $R$ 点的位置，其他方法的测量可作校正。

(2) Working from whole to part.

先整体后局部完成测量。

According to this principle, it is always desirable to carry out survey work from whole to part. This means, when an area is to be surveyed, first a system of control points is to be established covering the whole area with very high precision. Then minor details are located by less precise methods, as shown in Figure 14.2.

根据这一原理，整个区域的测量工作往往要求从整体到局部来完成。这意味着，当进行一个地区的测量时，应先建立覆盖整个地区的高精度控制点系统。然后对细小的局部测定可较粗略地完成，如图 14.2 所示。

The idea of working this way is to prevent the accumulation of errors and to control and localize minor errors which, otherwise, would expand to greater magnitudes if the reverse process is followed, thus making the work uncontrolled at the end.

这种方法有利于控制误差、减小误差的累积；否则若按先局部后整体的相反过程测量，则会使误差不断累积，导致最后测量工作不可控制。

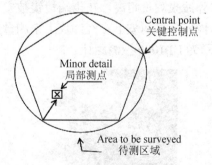

Figure 14.2　Working from Whole to Part

图 14.2　先整体后局部

## 14.4　Modern Surveying Instruments　现代测量仪器

In conventional surveying, chain and tape are used for making linear measurements while compass and ordinary theodolites are used for making angular measurements. Levelling work is carried out using a Dumpy level and a levelling staff. With such surveying instruments, survey work will be slow and tedious.

在传统的测量中，卷尺用来进行线性测量，而罗盘和普通经纬仪用于角度测量。水准测量是用一个老式水准仪和一个水准标尺进行工作的。这种测量仪器通常使得测量工作繁冗且效率极低。

Hence modern surveying instruments are becoming more popular and they are gradually replacing old surveying instruments such as compass and dumpy level. With modern surveying instruments, survey work will be precise, faster and less tedious. Some of the modern surveying instruments are discussed in brief, in the following pages.

因此，现代测量仪器变得更受欢迎，它们逐步取代了罗盘和老式水准仪等传统测量仪器。现代测量仪器使得测量工作更为精确、快捷及方便。部分现代测量仪器如下所述。

### 1. Electronic Distance Measurement (EDM) Instruments 电子测距(EDM)仪

Direct measurement of distances and their directions can be obtained by using electronic instruments that rely on propagation, reflection and reception of either light waves or radio waves. They may be broadly classified into three types:

距离和方位的量测可直接通过电子测距仪完成，它是通过光波和无线电波的传播、反射和接收完成测量。可以大致概括为三种类型：

1) Infrared wave instruments 红外测距仪

These instruments measure distances by using amplitude modulated infrared waves. At the end of the line, prisms mounted on target are used to reflect the waves. These instruments are light and economical and can be mounted on theodolites for angular measurements. The range of such an instrument will be 3km and the accuracy achieved is ± 10mm. Distomat DI 1000 is shown in Figure 14.3.

这种仪器是利用调幅红外波来测量距离。在末端棱镜被安装在目标处来反射光波。这种设备利用光电性能，轻质且经济，故可安装在经纬仪上来测量角度。最远工作距离为 3km，误差 ±10mm。如图 14.3 所示为 DI 1000 型测距计。

Figure 14.3　DISTOMAT DI 1000
图 14.3　DI 1000 型测距计

## Basic Concepts in Surveying

It is a very small, compact EDM, particularly useful in building construction and other civil engineering works, where distance measurements are less than 500 m. It is an EDM that makes the meaning tape redundant. To measure the distance, one has to simply point the instrument to the reflector, touch a key and read the result.

这是一种紧凑的电测仪，广泛应用在房屋建设和其他土建工程中。距离范围在 500 米以内。有此仪器，卷尺则毫无用处。测量距离时，只要简单地将设备指向反射装置，按键后就可阅读测量结果。

2) Light wave instruments  光波测距仪

These are the instruments which measures distances based on propagation of modulated light waves. The accuracy of such an instrument varies from 0.5 to 5 mm/ km distance and has a range of nearly 3 km. As shown in Figure 14.4.

光波测距仪是仪器基于调制光波的传播来测量距离。该仪器的精度范围为每千米 0.5～5mm，测量范围接近 3km。如图 14.4 所示。

This instrument which works based on the propagation of modulated light waves, was developed by E. Bergestand of the Swedish Geographical Survey. The instrument is more suitable for night time observations and requires a prism system at the end of the line for reflecting the waves.

该量测仪器的工作基于调制光波的传播，由瑞典地理勘测 E. Bergestand 部门研制。该仪器更适合夜间观察，在量测末端需设棱镜系统来反射光波。

Figure 14.4　Geodimeter
图 14.4　光波测距仪

3) Microwave instruments  微波测距仪

These instruments make use of high frequency radio waves. The instruments were invented as early as 1950 in South Africa by Dr. T.L. Wadley, as shown in Figure 14.5. The range of these instruments is up to 100 km and can be used both during day and night.

该测量仪器利用高频无线电波工作。这种仪器早在 1950 年由南非 T.L. 沃德利博士发明，如图 14.5 所示。该仪器的测量范围可达 100km，可用于白天和夜晚的测量。

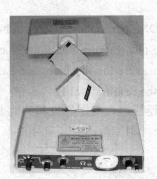

Figure 14.5　Tellurometer

图 14.5　微波测距仪

It is an EDM which uses high frequency radio waves (micro-waves) for measuring distances. It is an highly portable instrument and can be worked with 12 to 24 volt battery. For measuring distance, two Tellurometers are required, one to be stationed at each end of the line, with two highly skilled persons, to take observations. One instrument is used as a master unit and the other as a remote unit. Just by pressing a button a master can be converted in to remote unit and vice-versa. A speech facility (communication facility) is provided to each operator to interact during measurement.

这是一种采用 12～24 伏特的电池、使用高频无线电波(微电波)测量距离的可便携测距仪。测量距离时需要两个微波测距仪，待测距离的两端各放一个，需安排两个专业人员来采集数据。其中一台仪器是用来作为主要观测，另一台作为远程测量。通过按钮操控可完成两台仪器功能的互换。在进行测量时操作员可通过配备语音设备进行工作交流。

### 2. Total Station　全站仪

It is a light weight, compact and fully integrated electronic instrument combining the capability of an EDM and an angular measuring instrument such as wild theodolite, as shown in Figure 14.6. It can perform the following functions.

该仪器十分轻便，设计紧凑。集成了电子测距离(EDM)及角度测量的全部测量功能，如野外经纬仪。全站仪如图 14.6 所示。它具有如下功能。

Figure 14.6　Total Station

图 14.6　全站仪

Distance measurement; angular measurement; data processing; digital display of point details; storing data in an electronic field book.

距离测量、角度测量、数据处理、测点详细信息显示及在电子外业(野外作业)手簿中储存数据。

3. Global Positioning System  全球定位系统

This system is developed by U.S. Defense department and is called Navigational System with Time and Ranging Global Positioning System (NAVSTAR GPS) or simply GPS, as shown in Figure 14.7. For this purpose U.S. Air Force has stationed 24 satellites at an attitude of 20 200 km above the earth's surface. The satellites have been positioned in such a way, at least four satellites will be visible from any point on earth.

该系统是由美国国防部开发,被称为实时、全方位的全球定位系统(导航星测时与测距全球定位系统,NAVSTAR GPS),如图14.7所示。为此美国空军已经在距地球表面20 200公里处布置了24个人造卫星,这些卫星的定位使得在地球上任意地点均可见至少四颗这样的人造卫星。

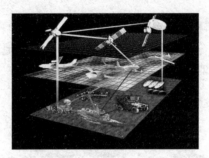

Figure 14.7  Global Positioning System
图 14.7  GPS 系统

The user needs a GPS receiver to locate the position of any point on ground. The receiver (as shown in Figure 14.8) processes the signals received from the satellite and compute the position (latitude and longitude) and elevation of a point with reference to datum.

用户需要有一个GPS接收装置来定位地面上任意一点的位置。接收装置(如图14.8所示)根据从卫星接收的信号来定位测点位置(经度和纬度)及其高程。

Figure 14.8  GPS Receiver
图 14.8  GPS 接收装置

## 14.5　GIS　GIS 系统

### 1. Concepts of GIS　GIS 的基本概念

GIS (Geographic Information System) consists of a set of computerized tools and procedures that can be used to effectively encode, store, retrieve, overlay, correlate, manipulate, analyse, query, display and disseminate land-related information.

GIS(地理信息系统)由一组计算机工具和程序组成，可以有效地进行编码、存储、反演、整合、关联、操作、分析、查询、成列和传播与地面相关信息的分析。

The main factor which separates GIS from other information storage and retrieval systems is the spatial and specific location features given in a co-ordinate form, for reference and use as an important variable in quantitative analysis. The spatial data is generally in the form of maps, which could be showing topography, geology, soil types, forest and vegetation, land use / land cover, water resources availability etc. and stored as layers in the digital form in the computer, as shown in Figure 14.9.

GIS 与其他信息存储和检索系统的区别在于它以一个协同方式给出了区域的空间特性，来作为重要的变量定量分析的参考。空间数据一般形成地图，可以显示地形、地质、土壤类型、森林和植被、土地利用/土地覆盖、水资源等信息，并在计算机中以数字层的形式存储，如图 14.9 所示。

Figure 14.9　GIS
图 14.9　GIS 系统

Besides spatial data, one could also have attribute data like statistics, texts and tables. New maps can be easily generated by integrating many layers of data in a computer. Thus, a GIS has a database of multiple information layers that can be manipulated or analyzed to evaluate relationships among the selected elements, from all that is available.

除了空间数据外，统计数据、文本信息及表格也十分有用。通过整合计算机中多层的数据可以很容易地形成新的地图。因此，GIS 有多个信息层的数据库，可以用来控制或分析评估所有待测部分之间的关系。

GIS can be looked as an integrating technology and a valuable tool for wide range of applications both in the public and private enterprises for storing events, explaining events, analyzing and predicting results and planning the strategies for making decisions before hand.

GIS 可以视为一种集成技术及有用的手段，用于公共和私人企业中存储、解释、分析预测结果，对将要实施的计划提前做好分析、规划。

2. GIS Applications　　GIS 应用

GIS is an effective tool for the application and monitoring the infrastructure of a town/city or a region. In addition, GIS is helping in the following applications.

GIS 是一种有效的应用工具，用于有效地监测一个小镇/城市或地区的基础设施。此外，GIS 可用于下述领域：

1) Project planning　　项目规划

It is helpful for detailed planning of a project having large spatial components.

可对大型空间组件的项目进行详细规划。

2) Proper decision　　恰当决策

GIS is capable of making automatic decisions based on study and analysis of data, made available to it.

GIS 能够基于研究和分析数据进行自动决策。

3) Generates map　　地图生成

GIS generates maps from the data available from data bases, as shown in Figure 14.10. Hence, GIS is also referred as "Mapping Software".

GIS 对数据库中的数据生成地图，如图 14.10 所示。因此，GIS 也被称为"地图软件"。

Figure 14.10　　Generate Map
图 14.10　　生成地图

4) Organizational integration　　系统整合

GIS helps in utilizing the data resources available to its maximum. The data available can be shared by different departments there by enhancing productivity and organizational efficiency.

GIS 可以最大限度地利用数据资源。不同部门可共享数据来提高工作效率和组织效率。

5) Valuable tool　实用的工具

GIS is a valuable tool in depicting calculations on volume of soil erosion due to floods and land slides, earth work quantities for engineering projects etc.

GIS 对于因洪水、沙土流失及工程施工造成的土壤侵蚀的分析是非常有用的。

## 14.6　Remote Sensing and Their Applications  遥感技术及其应用

Remote Sensing can be defined as obtaining information about an object by observing it from a distance and without coming into actual contact with it, e.g. Eye sight, Photos taken from a camera.

遥感可以实现不接触被测目标而通过远距离观察来获取其信息，如通过相机采拍的照片等。

In general, the word remote sensing is used for data collection from artificial satellites orbiting the earth and processing such data to make useful decisions. Hence, remote sensing can also be defined as the science and art of collecting information about an object, area or phenomenon from a far-off place without coming in contact with it. Remote sensing is shown in Figure 14.11.

一般来说，遥感是指从绕着地球运行的人造卫星收集数据，并通过处理这些数据来作出决策。因此，遥感也可以被定义为收集相关信息(如对无法接触的、物体、区域等)的一门学科。遥感过程如图 14.11 所示。

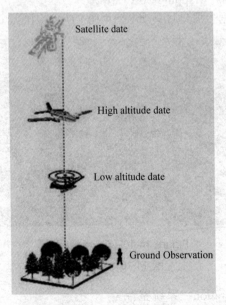

Figure 14.11　Remote Sensing
图 14.11　遥感过程

The satellite remote sensing system (as shown in Figure 14.12) consists of space segment, sensor system, ground segment.

卫星遥感系统(如图 14.12 所示)由空间观测部分、传感器、地面接收部分组成。

Figure 14.12　The Satellite Remote Sensing System
图 14.12　卫星遥感系统

1) Space Segment　空间观测部分

The space segment consists of the satellites orbiting the earth at a very high attitude, e.g. IRS-1A is launched to an attitude of 904 km and orbits the earth in North-South direction.

该部分由绕地球高空飞行的轨道卫星组成。例如，IRS-1A 在南北方向上被发射到距地 904 公里绕地球轨道运行。

2) Sensor System　传感器

Sensors are instruments which sense the objects on the surface of the earth and record them. The sensors mounted on satellites (called scanners) scans the objects on the surface of the earth. A set of consecutive scan lines forms an image, which is sent back to the earth receiving stations called ground segments.

传感器是感知在地球表面的对象并进行记录的仪器。安装在卫星(称为扫描仪)上的传感器自动记录地球表面的物体变化。一组连续的扫描线形成一个图像后发送回地球上的地面接收站。

3) Ground Segment　地面接收

They are the stations put up on the ground surface to receive the signals sent from remote sensing satellites. The signals received from the satellites are converted to images and stored in digital format. The images are to be processed to retrieve useful information. This process is called Image Processing, as shown in Figure 14.13.

这是建立地面的表面用以接收从遥感卫星发来的信号的基站。接收到的卫星信号被转换成图像并以数字形式存储。图像被加以处理以获取有用的信息。这一过程称为图像处理，如图 14.13 所示。

Figure 14.13　Image Processing
Figure 14.13　图像处理

# New Words and Expressions　生词和短语

1. surveying　　*n.* (土地)测量；考察
2. levelling　　*n.* 水准测量；校平
3. elevation　　*n.* 高地；海拔；提高；崇高；正面图
4. Plane surveying　　平面测量
5. Geodetic surveying　　大地测量
6. spheroidal　　*adj.* 类似球体的，球状的
7. triangles　　*n.* 三角形，三角形态(triangle 的复数形式)
8. parallel　　*n.* 平行线；对比
　　　　　　*adj.* 平行的；类似的，相同的
9. topographical　　*adj.* 地志的；地形学的(等于 topographic)
10. canal　　*n.* 运河；[地理] 水道；[建] 管道；灌溉水渠
11. hydrographic　　*adj.* 水道测量数的；水道学的
12. astronomical　　*adj.* 天文的，天文学的；极大的
13. reconnaissance　　*n.* [军] 侦察，勘测(等于 reconnoissance)；搜索；事先考查
14. dimension　　*n.* [数] 维；尺寸；次元；容积
15. magnitudes　　*n.* 大小；量级；[地震] 震级；重要；光度
16. linear　　*adj.* 线的，线型的；直线的，线状的；长度的
17. angular　　*adj.* [生物] 有角的；生硬的，笨拙的；瘦削的
18. infrared　　*n.* 红外线
19. microwave　　*n.* 微波
20. theodolite　　*n.* 经纬仪
21. tellurometer　　*n.* 微波测距仪；精密测地仪；无线电测距仪
22. panel　　*n.* 仪表板；嵌板
23. co-ordinate　　*n.* 坐标

24. curvature      n. 弯曲，[数] 曲率
25. refraction     n. 折射；折光
26. latitude       n. 纬度；界限；活动范围
27. longitude      n. [地理] 经度；经线
28. topography     n. 地势；地形学；地志
29. orbiting       v. [航][天] 轨道运行；轨道运动；转圈(orbit 的 ing 形式)
30. Sensor system       敏感元件系统
31. desertification     n. (土壤)荒漠化；沙漠化(等于 desertization)
32. tectonic map        大地构造图

# Exercises  练习

Ⅰ. Write a T in front of a statement if it is true according to the text and write an F if it is false.

1. In general, surveying deals with measurements in an horizontal plane while levelling deals with measurements in a vertical plane.

2. Plane Surveying is that type of surveying in which the mean surface of the earth is considered as a plane and the spheroidal shape is neglected.

3. The Engineering Survey is carried out to determine the absolute location of any point on the surface of earth.

4. In conventional surveying, chain and tape are used for making linear measurements while compass and ordinary theodolites are used for making angular measurements.

5. The accuracy of light wave instrument varies from 0.5 to 5 mm / km distance and has a range of nearly 2 km.

6. Remote Sensing can be defined as obtaining information about an object by observing it from a distance and coming into actual contact with it.

Ⅱ. Complete the following sentences.

1. The two primary divisions of survey are _____.
2. Please list five classifications of the surveying : _____.
3. International system of units (SI) is a form of absolute system of unit. In this the units of length, mass and time are _____.
4. Levelling work is carried out using _____.
5. Electronic Distance Measurement (EDM) Instruments may be broadly classified into three types:_____.
6. Total station can perform the following functions: _____.

Ⅲ. Translate the following terms and sentences into Chinese.

1. In other words, when small areas are to be surveyed, the curvature of the earth's surface is ignored and such a surveying is called plane surveying.

2. According to this principle, the relative position of a point to be surveyed should be located by measurement from at least two points of reference.

3. At the end of the line, prisms mounted on target are used to reflect the waves. These instruments are light and economical and can be mounted on theodolites for angular measurements.

4. The satellites have been positioned in such a way, at least four satellites will be visible from any point on earth.

5. Besides spatial data, one could also have attribute data like statistics, texts and tables.

6. The analysed and manipulated result of the data has to be displayed or presented to the user.

Ⅳ. Translate the following terms and sentences into English.

1. 测量可基于测量领域的性质、测量的物体和使用的仪器来分类的。
2. 在这种类型的测量中，信息是通过从选定的点拍照的方式收集的。
3. 力学的四个基本概念是长度、时间、质量和力。
4. 根据这一原则，开展从整体到局部的测量工作始终是可取的。
5. 用现代测量仪器测量，测量工作将是准确的、快速的和不乏味的。
6. 在测量过程中提供给每个操作员一种语音设备以进行互动。

# Answers   答案

Ⅰ.

1. T   2. T   3. F   4. T   5. F   6. F

Ⅱ.

1. Plane surveying; Geodetic surveying

2. Topographical Survey; Hydrographic Survey; Astronomical Survey; Engineering Survey; Aerial Survey

3. meter(m), kilogram (kg) and seconds(s)

4. a Dumpy level and a levelling staff

5. Infrared wave instruments; Light wave instruments; Micro wave instruments

6. Angular measurement; Data processing; Digital display of point details; Storing data is an electronic field book

# Chapter 14 Basic Concepts in Surveying

Ⅲ.

1. 换句话说，当进行小范围测量时，地球表面的曲率被忽略，这种测量称为平面测量。
2. 根据这一原则，被测量的点的相对位置应位于通过测量至少两个参考点确定的位置。
3. 在线的末端，安装在目标上的棱镜用于反射波。这些仪器又轻又经济，可以安装在用于角测量的经纬仪上。
4. 卫星以这样一种方式放置，从地球上的任何一点至少有四颗卫星是可见的。
5. 除了空间数据，也可以有属性数据，如统计、文本和表格。
6. 数据的分析和操作结果必须被展示或呈现给使用者。

Ⅳ.

1. Surveys may be classified based on the nature of the field of survey, object of survey and instruments used.
2. In this type of survey, information is collected by taking photographs from selected points.
3. Four fundamental concepts of mechanics are length, time, mass and force.
4. According to this principle, it is always desirable to carryout survey work from whole to part.
5. With modern surveying instruments, survey work will be precise, faster and less tedious.
6. A speech facility is provided to each operator to interact during measurement.

# Chapter 15   Disaster Prevention and Reduction
# 第 15 章   防灾与减灾工程

Every year, more than 200 million people are affected by droughts, floods, cyclones, tsunamis, earthquakes, wildfires and other disasters associated with natural hazards.

每年都有超过 200 万人受到干旱、洪水、飓风、海啸、地震、大火和其他自然灾害的影响。

Growing populations, environmental degradation and global warming are making the impacts worse, creating greater disasters and making the need to find better ways to protect people more urgent.

人口增长、环境恶化和全球变暖等加剧带来了更大的灾难，寻求更好的方法来保护人类自己变得十分紧迫。

## 15.1   Principal Causes of Disasters   灾害的主要成因

Disaster refers to all the occurrences that can cause destructive influence to the safety of people's life and property and the environment that they rely on to survive.

能够对人类生命安全、财产及赖以生存的生活环境造成破坏性影响的事件，称为灾害。

1) Natural Disasters   自然灾害
- Rain and wind storms;
- Floods;
- Biological agents (micro-organisms, insect or vermin infestation);
- Earthquakes;
- Volcanic eruptions.

如暴风雨、洪水、生物制剂(微生物，昆虫或寄生虫的大量滋生)及火山喷发等。

Generally disastrous incidents include the following four types: natural disaster, accident disaster, public health emergency and social security emergency. Among these, natural disaster includes flood, drought, meteorological disaster, geological disaster, forest fire and major biological disaster. The natural disasters in China are characterized by a large number of types, high regional and seasonal concentration, extensive territorial distribution, high occurrence frequency and heavy losses. The harms of the disasters mainly include:  casualty, economic

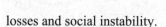

losses and social instability.

灾害通常有四种类型：自然灾害、事故灾难、公共安全和突发社会安全事件。其中，自然灾害包括洪水、干旱、气象灾害、地质灾害、森林火灾和重大生物灾害等。中国自然灾害的特点多种多样，地域、季节频发处集中发生，分布区域广，发生频率高，损失惨重。灾害造成的危害主要包括：伤亡、经济损失、社会不稳定。

2) Man-made Disasters 人祸

- Acts of war and terrorism;
- Fires;
- Water (broken pipes, leaking roofs, blocked drains, fire extinguishing);
- Explosions;
- Liquid chemical spills;
- Building deficiencies (structure, design, environment, maintenance);
- Power failures.

人祸包括战争和恐怖主义、火灾、水灾(水管破裂、排水受阻、屋顶渗漏、消防灭火)、爆炸、液体化学品泄漏、建筑物破坏(结构、设计、环境、维护)、停电。

Natural disasters cannot be prevented, but measures can be taken to eliminate or reduce the possibility of trouble. Regardless of the many forms a disaster may take, the actual damage to collections is usually caused by fire or water. Even when they are not the initial factor, fires and floods almost invariably occur as secondary causes of library and archives disasters.

人类无法阻止自然灾害的发生，但可以采取措施来消除或减少其产生的影响。无论灾难以何种形式发生，通常是由火或水所带来的损害。即使它们不是灾害的起因，火灾和水灾都会作为次生灾害而发生。

Disaster prevention and reduction refer to the processes to understand the reasons, impacts and consequences of these disasters, carry out prevention, preparation and exercises in advance, conduct daily disaster prevention and avoidance properly and correctly, and reduce loss and impact to the minimum at the time of disaster.

防灾减灾是指了解这些灾害的原因、影响和后果的整个过程，提前预防准备和练习，适当并正确地开展日常的灾害预防防治措施，当在灾害发生时，可使损失减少并将影响降至最低。

## 15.2 Earthquake 地震

Earthquake refers to the vibration of the surface of the earth caused by the sudden release of the energy slowly accumulated in the interior of the earth. China is a country suffering frequent earthquakes. There have been many strong earthquakes in its history, such as the Hebei Tangshan Earthquake in 1976 and the Wenchuan Earthquake in 2008.

地震是指因地球内部缓慢积累的能量突然释放而引起的地球表面的震动。中国是一个地震多发的国家。历史上已有很多次强烈的地震，如 1976 河北唐山大地震，2008 年的汶川大地震。

## 1. Earthquake Forecasting  地震预报

1) Abnormalities of Underground Water  地下水异常

Springs seem muddy when it has not rained. Water flows out from the well when there has not been any wet weather. Water level rises or drops sharply. Bubbles come out of water. Some water changes its color. Some changes its taste.

没有下雨但是泉水变得浑浊；并非雨季，但水从井里溢出；水位大幅上升或下降；水中冒泡；水的颜色或味道改变。

2) Abnormalities of Biology  生物异常

Mule, horse, cattle and sheep are not willing to enter or stay in their stables and pens. Ducks refuse to get into water. Dogs bark madly. Mice and rats run away from houses. Pigeons fly in panic and refuse to return to their nest. Snakes come out of their shelters in winter. Fishes jump in panic out of water. As shown in Figure 15.1. Besides, some plants also show abnormalities before an earthquake. They may sprout, blossom, grow fruits, wither in large areas and become greatly thriving in an unduly season.

骡、马、牛和羊不进圈；鸭子不愿入水；狗吠；老鼠纷纷跑离房屋；鸽子狂飞，不愿回窝；冬眠的蛇爬出巢穴；鱼惊恐地跳出水面，如图 15.1 所示。此外，有些植物在地震前也显示出异常的变化，可能会突然发芽、开花、结果，大范围地枯萎，或在不适当的季节里异常茂盛。

Figure 15.1  Abnormalities of Biology

图 15.1  生物异常

3) Earth Light and Earth Sound  地光与地声

Moments before an earthquake, loud earth sound can be heard and strong earth light can be seen.

地震前，可以听到大地发出很大的声音，并可看见大地发出很亮的光芒。

## 2. Regular Earthquake Emergency Preparation  常规地震应急准备

Take safety and reinforcement measures inside the house and the school classrooms. For example, secure lockers to prevent them from falling down. Paste transparent films and tapes on windowpanes. Place heavy objects in lower places.

# Chapter 15 Disaster Prevention and Reduction

在屋内和学校教室里采取安全加固措施。例如，设置安全锁柜，以防止它们掉落。在窗玻璃上粘贴透明薄膜和胶带。将重物放置在较低的地方。

Prepare earthquake kits (including drinking water, food, and medicine for daily use, raincoat, electric torch and radio), as well as materials to build tents.

准备地震应急包(包括饮用水、食物和日常使用的药品、雨衣、手电筒和收音机)以及搭建帐篷的材料。

Learn the commonsense knowledge about earthquake portents and other things and also the knowledge and skills for self rescue and mutual rescue.

掌握有关地震方面的一些常识及自救和互救的知识技能。

## 15.3 Geological Disasters: Landslide, Collapse and Debris Flow 地质灾害：滑坡，崩塌和泥石流

### 1. Landslide, Collapse and Debris Flow 滑坡，崩塌和泥石流

Landslide refers to the geological disaster that the soil or rock body on a slope slides down such slope, in a whole or in several parts, due to gravity under the influence of river erosion, underground water activity, earthquake and artificial slope cutting.

滑坡是指斜坡整体或部分因其重力、河流侵蚀，地下水活动，在地震和人工削坡的作用下，土壤或岩石体在斜坡上向下滑动的地质灾害。

Collapse refers to the geological disaster that the soil or rock body on a steeper slope suddenly separates from the mountain due to gravity and falls to rolls down and piles up at the foot of the slope or in a valley.

崩塌是指土壤或岩石体在更为陡峭的山坡上，由于重力的作用，突然与山体分离滚落并堆积在坡脚或山谷中的地质灾害。

Debris flow refers to the geological disaster frequent in mountainous regions that floods flow along a slope with large amounts of mud and rocks. Debris flows are usually abrupt, ferocious, fast-moving and short in time. Having the destructive powers of both collapse and landslide and flooding, they cause far more severe damages than collapse, landslide or flood alone.

泥石流是山区频发的地质灾害，大量的泥土和岩石由于洪水的作用，沿着斜坡流动。泥石流通常匆匆来袭，来势汹汹，短时间内转移速度快。同时具有塌陷、山体滑坡和洪水的破坏力，可以引起比崩塌、滑坡或洪水更为严重的损害。

### 2. Portents of Landslide. Collapse and Debris Flow 滑坡、倒塌及泥石流征兆

1) Portents of Landslide 滑坡征兆

Before a major landslide, abnormalities will occur along the slope foot in the front of the sliding slope: that springs dead from many years start to have water flowing, that springs (or wells) suddenly dry up, that the water level in a well (or a drilled hole) changes sharply, and that the water in springs or wells turn turbid.

重大山体滑坡前常有异常发生：在坡脚缓慢滑坡中干涸多年的沟渠开始有水流淌，沟渠(或水井)突然干涸，水井(或钻孔)中的水位有明显的变化，并且泉水或井水变得浑浊。

Horizontal and vertical radial fissures appear in the front of the landslide body. Bumps (or uplifts) appear on the soil along the slope foot in the front of the sliding slope. There are sounds showing rocks cracking, sheared or pressed.

在滑坡体的前部出现水平和垂直的径向裂缝。沿着坡脚，在滑动边坡前的土壤上出现凸块(或隆起)。发出的声音显示岩石在开裂、剪切或受压破坏。

Just before the landslide begins, the rock and soil around the landslide will have minor collapse and loosening. The crack at the back of the landslide expands drastically. Hot vapor or cold wind blows out of the crack.

在滑坡刚刚开始前，滑坡周围的岩石土壤将有轻微的塌陷和松动。山体后面的滑坡裂缝急剧扩大。热蒸汽或寒风从裂缝中散出。

Just before the landslide begins, animals within the scope of the landslide will fall in great panic. For example, pigs, dogs and cattle will become restless and scared and will not go to sleep. Mice and rates will run around, not going into their shelters.

在滑坡刚刚开始前，滑坡范围内的动物会陷入极大的恐慌。例如，猪、狗和牛会变得焦躁不安和害怕；老鼠会四处乱跑，而不再进住所。

2) Portents of Collapse  倒塌的征兆

Falling rocks and soils and minor collapse constantly occur at the front edge of the collapsing mountain body.

在坍塌山体的前缘，岩石和土壤的掉落及轻微塌陷会不断地发生。

New breaks and cracks appear at the foot of the collapse. Sometimes sounds showing rocks torn and rubbing can be heard.

在塌陷的底部出现新的断裂和裂缝。有时能听到岩石断裂和摩擦的声音。

There are abnormalities of heat, air, underground water and animal.

气温、空气、地下水和动物会有异常反应。

3) Portents of Debris Flow  泥石流的征兆

Be alert when there is rainstorm or continual rainfall.

警惕暴雨或持续降雨。

Water flow suddenly discontinues in a river or a flood suddenly rises with large amounts of floating grasses and trees.

水流量突然在河道中终止或水位突然上升，并有大量的浮草和树木。

It suddenly turns dark deep in a valley with big roaring sound or slight ground shaking.

山谷中天突然变暗，并伴之巨大的轰鸣声，或地面轻微地晃动。

3. Relevant Disaster Prevention and Rescue Knowledge  防灾及救援知识

1) Landslide Prevention and Rescue(See Figure 15.2)  滑坡预防和救援(见图15.2)

(1) How to escape at the time of landslide in a mountain:

在山体滑坡发生时如何逃生：

# Disaster Prevention and Reduction

## Chapter 15

Figure 15.2 Disaster Prevention and Rescue

图 15.2 灾害预防和救援

Run in the direction vertical to that in which the rocks are rolling when you find yourself in a landslide in a mountain.

当你发现自己在滑坡山体上时，沿岩石滚动的垂直方向逃跑。

Suggestions: ① Hide under a strong obstruction or crouch in a ditch or trench on the ground.

建议：躲在坚硬的障碍物下或蹲在地面上的沟渠或沟槽内。

② Protect your head well and use your clothes to wrap your head up at time of landslide.

尽可能地保护好你的头部，并用在滑坡发生时用你的衣服包住头部。

What to Avoid:

要避免：

① Run downhill in the direction in which the rocks are rolling.

沿着岩石滚动的方向逃跑。

② Not protect your head.

不保护自己的头部。

(2) Landslides are likely to happen:

山体滑坡可能发生的时间：

① after a heavy rain or in a period of continual rains.

大雨或持续降雨后

② during various kinds of building construction and the earthquake period.

各类建筑施工和地震期间。

③ during the snowmelt period in the spring each year.

每年春季的融雪期间。

(3) How to choose a house in a region susceptible to landslides:

在易受滑坡影响的地区如何选择房子：

It is important to examine the changes occurred to the house and the things around it.

检查房子和周围事物发生的变化是非常重要的。

Key Points to Self Rescue and Mutual Rescue:

自救和互救的要点：

① Check if the power line poles around the house incline toward one direction.

检查房子周围是否有倾斜的电线杆。

② Check if the asphalt roads around the house have deformed.

检查房子周围是否有已经变形的沥青路面。

What to Avoid:

应避免：

Mistake the cracks on doors and walls and the inclination of power line poles caused by artificial factors as portents of landslide.

误以为在门和墙裂缝以及倾斜的电线杆等山体滑坡的征兆是由人为因素所造成的。

(4) How to check a usable house after a landslide:

山体滑坡后如何检查房屋的安全性：

Check carefully if any facilities of the house have been damaged.

仔细检查房子中已损坏的设施。

Key Points to Self Rescue and Mutual Rescue:

自救和互救的要点：

Before moving into the house, check carefully if the facilities in the house such as water, electricity and gas have been damaged and if pipelines and power lines have been broken or torn apart. If any trouble is located, call for repair immediately.

在搬进房子之前，仔细检查房子的水、电、气等设施是否遭到严重的破坏；检查管道和电源线是否局部破坏或撕裂。如果发现有任何问题，立即找人修理。

What to Avoid: Move in without carefully checking the safety of water, electricity and gas.

应避免：在没有仔细检查水、电、气的安全前就进入房屋。

(5) What to do when you pass an area susceptible to landslide by bicycle:

骑自行车通过容易滑坡的地区时该怎么办：

Observe carefully and pay attention to your safety in riding.

仔细观察，并在骑车时注意安全。

Key Points to Self Rescue and Mutual Rescue:

自救和互救的要点：

① Pay attention to various dangers that may appear on the road, such as falling rocks or branches at any time.

要注意道路上可能出现的各种危险，如随时可能出现的落石或树枝。

② Check if there are collapses and ditches on the road in front of you to avoid danger.

检查面前的道路是否有沟渠和崩塌，以免发生危险。

What to Avoid:

应避免：

① Ride by without checking.

未经仔细观察就通过该区。

② Ride past a region immediately after a landslide.

山体滑坡后就立即骑过该区。

(6) What to do after a landslide has struck:

遭到山体滑坡后该做什么：

# Disaster Prevention and Reduction

Do not enter the region the landslide has struck to look for lost properties.
不要为了挽救财物贸然进入滑坡区域。

Key Points to Self Rescue and Mutual Rescue:
自救和互救的要点：

① Participate in the rescue of other victims of the landslide.
参加救援其他受到滑坡影响的受害者

② Do not return to the area struck by the landslide until the landslide danger period has passed so as to avoid being caught in a second landslide.
在滑坡危险期过去前不要返回该地区，以避免遭受第二次滑坡的伤害。

③ Enter your house only when you have confirmed that your house is safe and good, if the landslide has passed and your house is far away from it.
当滑坡危险已过去，并且房子离它很远时，只有确认房屋安全时方可进入。

What to Avoid:
应避免：

① You should not go home to check the situation immediately after the landslide stops.
不应在滑坡停止后就立刻回家检查情况。

② To go home carelessly and neglect the possibility of continual landslides may get you endangered by a second landslide.
盲目回家并忽视可能的持续滑坡，会使自己遭遇第二次滑坡危险。

(7) What to do when you are on the sliding mountain part at the time of a landslide:
若恰在滑动的山体上时，该怎么做：

Keep calm without panic.
保持冷静而不惊慌。

Key Points to Self Rescue and Mutual Rescue:
自救和互救的要点：

① Run in the direction vertical to that of the landslide to escape and find a safe place around as quickly as possible.
沿滑坡垂直方向逃跑，尽快在周围找一个安全的地方。

② If it is impossible to continue to escape, quickly hold tight the fixed objects around you, for example, trees.
如果不能继续逃跑，迅速抓住你身边的固定物体，如树木。

What to Avoid:
应避免：

① Run to escape in the direction of the landslide.
向山体滑坡的方向逃离。

② Become scared and confused and roll down with the landslide.
害怕迷茫，并与滑坡山体一起滚下来。

(8) What to do when you are in an area not affected by an occurring landslide:
若地处不受滑坡影响的区域时该怎么办：

Report the situation of the disaster to relevant government departments or other relevant entities.

立即向相关政府部门或其他相关实体报告灾情。

Key Points in Self Rescue and Mutual Rescue:

自救和互救的要点：

① Do not be scared. Report detailed situation of the disaster to relevant government departments or other relevant entities as quickly as possible.

不要害怕，尽快向相关政府部门或其他相关实体详细报告灾情。

② Take good measures to protect your own safety.

采取良好的保护措施来保证自身安全。

What to Avoid:

应避免：

① Consider the disaster as not concerning you and do not report it.

认为灾难与自己无关而不向有关方面报告。

② Go to disaster rescue alone.

单独抢险救灾。

(9) How to select a temporary disaster shelter:

如何选择一个临时避难所：

Move to a safe location in advance which is the best way to defend against a landslide.

抵御山体滑坡最好的方式是提前转移到一个安全的位置。

Key Points to Self Rescue and Mutual Rescue:

自救和互救的要点：

① Select several safe shelters around your place in advance.

提前在周围选择几个安全的庇护所。

② Select disaster shelters located outside the two sides of the area susceptible to landslides. Under the condition of ensuring safety, select locations as far as possible from the place you live, with transportation, water and electricity supply as convenient as possible.

选择位于易受滑坡影响区外的避难所。在确保安全的情况下，选择离你的居住地尽可能远的位置，并尽可能方便运输和水、电的供应。

What to Avoid:

应避免：

① Select shelters up or down the area susceptible to landslide.

选择滑坡上面或下面作为避难处。

② Move from one dangerous area to another due to insufficient and incomplete investigation and examination.

在没有完整和及时地检查前从一个危险区转移到另一个危险区。

(10) How to select an evacuation route:

如何选择疏散路线：

You must determine a right evacuation route through field survey.

# Disaster Prevention and Reduction

通过实地调查，你必须确定一个合适的疏散路线。

Key Points to Self Rescue and Mutual Rescue:
自救和互救的要点：

① Invite experienced experts to conduct field survey.
邀请有经验的专家进行实地调查。

② Select a speedy, convenient and safe route.
选择一个快捷、方便、安全的路线。

What to Avoid:
应避免：

① Be too scared to choose the right route and enter dangerous areas.
因太害怕，而不能选择合适的路线，以致进入危险区域。

② Disobey the uniform arrangements of evacuation organizers and select your own evacuation route.
违背疏散组织者的统一安排，而选择自己的疏散路线。

(11) What to pay attention to when rescuing people and things buried in a landslide?
从两侧的山体滑坡处救助被埋在山体中的人与物品时应注意什么？

Key Points to Self Rescue and Mutual Rescue:
自救和互救的要点：

① Drain the water at the back of the landslide.
在滑坡后面沥干水。

② Start digging from the sides of the landslide.
从山体滑坡的侧面开始挖。

③ Save people first and then things.
救人第一，然后才是抢救物资。

What to Avoid:
应避免：

Dig from the lower end of the landslide and make the landslide move faster.
从山体滑坡的下端开挖，使滑坡移动速度更快。

2) Collapse Prevention and Rescue Knowledge　塌方的预防与救援知识

Steeper slopes or block roads, are susceptible to collapses, which often disrupt cause damaged or fall, or turn buildings at the foot of the slopes to be into debris flow directly if there is a flood. The natural dam formed when a collapse blocks and disrupt a river will generate backwater in the upstream, which may cause the river to overflow and result in flooding.

陡峭的斜坡或阻塞的道路很容易坍塌，并经常造成破坏或下陷。如果有泥石流，坡脚下的建筑会变得粉碎。坍塌阻塞会扰乱河流的流动，形成天然堤坝，使上游回水，这可能会使河水溢出并最终导致洪涝灾害。

3) Debris Flow Prevention and Rescue Knowledge　泥石流的预防和救援知识

(1) What to do when a debris flow occurs?
当泥石流发生时该怎么办？

When in the debris flow area, quickly escape from the two sides of the debris flow ditch. Do not run up or down the ditch(See Figure 15.3).

在泥石流发生时,应迅速逃离泥石流沟的双侧。不要沿着沟向上或向下逃跑(见图15.3)。

Figure 15.3  How to Escape?
图 15.3  如何逃生?

When outside the debris flow area, immediately report the situation to village (community), township (town) or industrial and mining enterprises located downstream the debris flow ditch that may be affected.

在泥石流发生区外,应立即将情况上报到村(社区)、乡(镇)或工矿企业等位于泥石流沟下游可能受影响的区域。

Pay close attention to the secondary disaster or even tertiary disaster on certain lifeline projects, such as reservoir, railway, highway, power plant, communication facility, and canal and channel, that the debris flow may cause, such as fire, flood, transportation disruption, explosion or building collapse.

密切关注泥石流的次生灾害,甚至对某些生命线工程产生的三级灾害,如对水库、铁路、公路、电厂、通信设施以及运河和沟渠等产生影响,造成火灾、洪涝、交通中断、爆炸或建筑物倒塌。

(2) How to escape when encountering a debris flow:

当遇到泥石流时如何逃生：

When encountering heavy rain as you are walking on foot along a valley, quickly move to safe higher places nearby. The more far away you are from the valley the better. Do not stay at the bottom of the valley for longer time.

当你沿着山谷徒步行走并遇到大雨时,快速移动到附近更高的安全地方。离山谷越远越好。不要长时间地停留在谷底。

Pay close attention to your surroundings; particularly listen carefully to whether there are thundering sounds from the distance in the valley. If there are, be in high alert. This might be the portent of a coming debris flow.

密切关注你周围的环境,特别是应仔细倾听是否有雷鸣声从远处山谷传来。如果有,应保持高度警觉的状态。这可能是即将到来的泥石流的征兆。

Select flat higher places as camp base. Avoid as much as possible the place below a slope with rolling rocks and large amounts of deposits. Do not camp at the bottom of a valley or on the riverbed of a stream.

选择较高的平坦的场地建造基地。尽可能地避免在斜坡下方滚动的岩石和大量淤积的地方。不要在谷底或河床上设营。

When a debris flow occurs, climb upwards immediately to the slopes on the two sides of the debris of flow in a direction vertical to that of the debris of flow. The higher you climb the better. The faster you escape the better. Never walk downstream the debris flow.

当泥石流发生时，立即朝着泥石流垂直的方向向上攀爬到泥石流两边的斜坡上。你爬得越高越好。你逃脱得越快越好。不要往泥石流下游走。

(3) Self rescue and mutual rescue in the disaster of debris flow:

泥石流灾难中的自救和互救：

Portents of a coming debris flow: A debris flow is likely to occur after long continual rainfall. The thundering sounds heard from a valley after heavy rains portends a coming debris flow.

泥石流即将到来的征兆：经长时间的持续降雨后，泥石流可能会发生。大雨过后听到雷鸣声从山谷传来，预示着泥石流即将到来。

## 15.4　Rainstorm and Flooding　暴雨和洪水

1. Rainstorm and Flooding(See Figure 15.4)　暴雨和洪水(见图 15.4)

Rainstorm refers to rains of huge precipitation. It often causes flooding and serious soil and water losses, which in turn causes infrastructure accidents, dam failure and crop flooding Particularly, in some low-lying areas and those on enclosed terrains, rainstorm may cause even more disasters.

暴雨是指巨大的降水降雨。这往往会导致洪水和严重的水土流失，从而发生基础设施损坏的事故，并对堤坝和农作物产生巨大的影响，特别是在一些低洼封闭地区，暴雨可能会造成更多的灾害。

Flooding mainly refers to the disaster that, when heavy rain concentrating within a short period, or long time rainfall, occurs within a river basin, the runoff accumulated in the river course exceeds its flood carrying capacity so that the river overflows its banks or causes dam failure and floods. Flooding is often resulted from the floods in rivers, lakes and reservoirs hit by rainstorms or from submarine earthquakes, hurricanes and dam collapses.

洪水主要是指短时间内大量地集中降雨或长时间降雨时流域内发生的灾难，在河道的累计径流超过其洪水的承载能力，导致河水泛滥，河道或大坝失效而决堤。洪水往往是河、湖、水库因暴雨或海底地震、飓风袭击和大坝坍塌造成的。

Figure 15.4　Rainstorm and Flooding
图 15.4　暴雨和洪水

2. Prevention and Control of Rainstorm and Flooding　暴雨和洪水的预防和控制

1) Rainstorm Prevention and Control　暴雨预防和控制

(1) Disaster portents of rainstorms:
暴雨灾害的征兆:
When intense precipitation concentrates in a river basin within a short period of time, prepare to prevent possible disasters of floods in mountainous regions, landslides and debris flows.
当区域强降水集中在一个较短的时间内, 在山区应准备预防洪灾带来的山体滑坡和泥石流等灾害。

(2) Disaster forecasting for rainstorms:
暴雨灾害预报:
Check the drainage system in crop fields and fish ponds. Prepare to drain flooded fields.
检查农田和鱼塘的排水系统, 准备从受淹田地里排水。
Cut off dangerous outdoor power supply in low-lying areas. Suspend outdoor operations in open spaces. Evacuate people in dangerous places and residents of perilous buildings to safe locations to take shelter from the rainstorm.
在低洼地区, 切断危险的室外电源。在空旷处暂停室外作业。将那些在危险场所和危险的建筑物中的居民转移到安全地点, 暂避暴雨。
Organizations located in dangerous places must stop operation of teaching and business. Take special measures to protect the safety of the students and children in schools and kindergartens and staff members.
处于险境的单位的教学和业务操作必须停止。采取措施, 以保护在学校和幼儿园里的学生、儿童及工作人员的安全。

2) Flooding Prevention and Control:
洪灾预防和控制:

(1) Disaster portents of flooding:
洪水灾害征兆:
Be alert when there are constant heavy rains and storms and the water level rises.
警惕持续强降雨和风暴时的水位上升

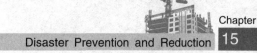

## Disaster Prevention and Reduction

(2) Disaster forecasting for flooding:
洪水灾害预报：

Pay close attention to the developments of weather condition, rainfall and flooding. Report timely to the flood control command offices and relevant authorities. Publicize the disaster early warning information to the people in the endangered area by the various means of television, radio, SMS message and drum and gong signals in a timely manner. Check for flood alarms frequently in the rainy season.

密切关注天气状况、降雨和洪水的发展情况。及时向防汛指挥办公室和有关部门汇报。向受灾区的居民通过各种手段如电视、广播、短信、锣鼓等及时公布灾害预警信息。在雨季时经常检查洪水警戒线。

3. Commonsense Knowledge on Disaster Prevention and Rescue in Rainstorm and Flooding
暴雨洪水灾害自救的常识

To prevent minor waterlogging inside a house, place water baffles or pile up earth ridges in front of the door according to local conditions. People living in bungalows should check the house and repair its roof before the rainy season comes.

为了防止房子轻微内涝，因地制宜地在屋内放置挡水板或在门前堆放土。住在平房的居民应该在雨季来临之前检查房屋和修复屋顶。

Cut off power supply immediately when water comes into the house from outside to prevent the electricity carried by the water from the power lines from causing harm to people.

当水从外部进入房子，应立刻切断电源，以防止电源线漏出的电对人造成伤害。

Observe carefully when walking in the water outdoors. Walk by the building walls so as to avoid falling into inspection wells and pits on the ground.

在户外的水中行走时应仔细观察。沿着建筑墙体行走，以避免落入坑井和路面的维修区中。

Get indoors to take shelter immediately when encountering thunderstorm and gale on the street. Do not throw garbage and litter into sewerage openings on the road to avoid causing blockage and flooding.

当在街上遇到雷雨大风时应立即到室内暂避。不要将垃圾乱扔到道路上开口的下水道内，以避免造成堵塞和洪水。

Watch for floods when in the mountain or in the open. Be alert particularly when the water coming from the upstream suddenly turns turbid or the water level rises quickly in a river. If walking by a lake or a river, crouch down at a lower place, put two legs together and two arms around the knees, press the chest against the knees, and lower your head as much as possible.

在山上或户外时，应留意洪水和河中的警报信号，尤其是当来自上游的水突然变浑浊或水位很快上升时。如果是路过湖泊或河流，应在一个较低的地方蹲下，两只胳膊抱紧两条腿，胸部紧贴膝盖，并尽可能低地低下头。

When riding a motorcycle or a bicycle, try finding another way to pass if the water on the road or under an overpass gets too deep. Do not try to pass these locations.

在骑摩托车或自行车时，如果在道路上或立交桥下的水变得太深，试着找一条路通过。

不要试图从这些地方通过。

What to Avoid: Never touch the ground with hands, take shelter from thunderstorm under trees, get close to water surface, or use cell phone.

避免事项：切勿用双手触碰地面、在树下躲避雷雨、靠近水面或使用手机。

## 15.5　Disaster Self Rescue　灾难发生后的自救

If you are buried under the ruins after the disaster, surrounded by complete darkness, and shut in an extremely small space, do not be scared, and keep calm(See Figure 15.5). Protect yourself by all means. Build the confidence to survive and believe that other people will protect you.

如果灾难后被掩埋，周围一片黑暗，并被困在一个非常狭小的空间里，不要害怕，要保持冷静(见图 15.5)。用各种手段保护自己。树立生存下去的信心，并相信其他人会来保护你。

Figure 15.5　Disaster Self Rescue
图 15.5　灾难自救

Keep your breath smooth. Remove the things on your head and chest. Cover your mouth and nose with wet clothes or other things when you smell coal gas or poisonous gas.

保持呼吸顺畅。移走头部和胸部周围的东西。当你闻到煤气或有毒气体时，用湿衣服或其他东西捂住嘴和鼻子。

Stay away from unstable collapsed things and other objects that might fall down above your body.

远离不稳定的倒塌物，因为它们有可能会砸到身体。

Expand and stabilize your survival space. Support broken walls with bricks and wooden sticks to prevent the environment from worsening after the occurrence of other secondary disasters.

扩大和稳定生存空间。用砖头和木棍支撑墙壁，防止因次生灾害导致的环境恶化。

Try your best to escape from the dangerous situation. If you cannot find a way to escape, preserve as much as energy as possible. Send signals for help to the outside by knocking on things that can make sound. Do not cry, scream, become agitated, or act blindly. Control your emotions or close your eyes to have some rest. Wait for the rescue personnel.

请尽量逃离危险的地方。如果你找不到逃生的路线，那就尽可能多地保存能量。通过敲东西制造声音来向外界发送求救信号。不要哭、尖叫、烦躁或盲目行事。控制自己的情绪，闭上你的眼睛休息。等待救援人员的到来。

If you are injured, try to band the wound to prevent bleeding too much.

如果你受伤了，尝试将伤口绑扎以防止出血过多。

Try to survive. If you have been buried under the ruins for a long time and the rescue personnel have not come or heard your signals for help, try your best to survive. Save water and food. Try your best to find food and drinking water. Your own urine can be used to satisfy your thirst when necessary.

试着生存。如果你已经被埋在废墟下很长一段时间，而救援人员没有来或听到你的求救信号，尽你所能生存下来。节约水和食物。请尽量寻找食物和饮用水。必要时，可以用自己的尿解渴。

## 15.6　Some Major Effects of Disasters
　　　一些主要的灾害影响

### 1. Fire　火

Books burn fairly slowly. Paper chars and crumbles when handled. Smoke and soot discolour books not otherwise affected. Microforms and audio-visual materials can be completely destroyed or damaged beyond repair.

书籍的燃烧相当缓慢。纸烧焦后再移动会变成粉末。烟尘会使书掉色。缩微胶卷和视听材料可以被完全摧毁或损坏，并且无法修复。

### 2. Water　水

Paper absorbs water at different rates depending on the age, condition and composition of the material. Generally speaking, books and manuscripts dated earlier than 1840s absorb water to an average of 80% of their original weight. Modern books, other than those made of the most brittle paper, absorb to an average of 60 % of their original weight.

纸张吸水速率因年代、环境和材料的组合而不同。一般来说，早于19世纪40年代的书籍和手稿可以平均吸收其原始重量80%的水分。除了由最脆的纸张制作而成的书籍，现代书平均能吸收其原始重量60%的水分。

Leather and parchment warp, wrinkle or shrink. The damage done to book covers may be irreparable. Water can cause gelatinization on parchment. After floods, mould rapidly begins to form in damp conditions. Audio-visual materials, photographs, microforms, magnetic media and other disks, are also vulnerable to water, and the damage depends on the type of the material, the length of exposure to water, its temperature, etc.

皮革和羊皮纸扭曲后会起皱或收缩。书封面上的损害可能是无法弥补的。水可以使羊皮纸糊化。洪水后，霉菌会在潮湿的条件下快速形成。视听材料、照片、缩微胶卷、磁介质、磁盘也容易受到水的损害，而损害的轻重取决于材料的种类、暴露在水中时间的长短和温度等。

### 3. Earthquakes 地震

Shelving may collapse and the contents be thrown on to the floor. Few books can withstand such treatment. Fire and water damage often result from seismic activity.

货架可能会坍塌，并且货物会掉落至地板上。很少有书可以承受得住这样的晃动。地震活动往往会导致火灾和水灾。

## 15.7 Disaster Plan 灾难应变计划

This usually involves four phases: Prevention; Preparedness; Response; Recovery.
这通常包括四个阶段：预防、准备、响应、恢复。

The following guide to producing a disaster plan outlines recommended action in all four phases, but prevention is the best protection against disaster, natural or man-made.
以下是救灾计划概述中四个阶段的所有建议行动指导，但预防能最好地防范自然的或人为的灾害。

Phase Ⅰ: Prevention
第一阶段：预防

- Identify and minimize the risks posed by the building, its equipment and fittings, and the natural hazards of the area.
  识别并减少建设、设备配件和受灾区所带来的危险。

- Carry out a building inspection and alter factors which constitute a potential hazard.
  开展验楼并消减潜在的危险因素。

- Establish routine housekeeping and maintenance measures to withstand disaster in buildings and surrounding areas.
  建立日常家务、建筑物和周边地区承受灾害的维护措施。

- Install automatic fire detection and extinguishing systems, and water-sensing alarms.
  安装自动火灾探测和灭火系统、水感应报警器。

- Take special precautions during unusual periods of increased risk, such as building renovation.
  因风险增加，在不寻常的时期采取特殊的预防措施，如建筑装修。

- Make special arrangements to ensure the safety of library or archival material when exhibited.
  展出时应作出特别安排，以确保图书或档案材料的安全。

- Provide security copies of vital records such as collection inventories, and store these off-site.
  提供重要记录的安全副本，如收集库存、异地储存。

- Protect computers and data through provision of uninterrupted power supply.
  通过提供不间断电源来保护计算机和数据。

# Disaster Prevention and Reduction

- Have comprehensive insurance for the library or archives, its contents, the cost of salvage operations, and potential replacement, re-binding and restoration of damaged materials.
  对图书馆或档案内容、救助作业成本、潜在变换、重新绑定和损坏材料的修复投保全面保险。

Phase Ⅱ: Preparedness
第二阶段：准备

- Getting ready to cope.
  准备好应对。
- Develop a written preparedness, response and recovery plan.
  制订一份书面的准备、回应和恢复计划。
- Keep the plan up-to-date, and test it.
  计划保持最新，并时时需要测试。
- Keep together supplies and equipment required in a disaster and maintain them.
  同时保持和维护灾难所需的物资和设备处于良好状态。
- Establish and train an in-house disaster response team. Training in :
  建立和培养一个内部的灾难响应救助队。培训内容有：
  - Disaster response techniques.
    灾害应急技术。
  - Identification and marking on floor-plans and enclosures of irreplaceable and important material for priority salvage.
    为了优先救助，应在地板围墙上、不可替代和重要的材料上做识别的记号。
- Prepare and keep an up-to-date set of documentation including :
  准备和保存最新的文档集，包括：
  - Building floor-plans, with locations of cut-off switches and valves.
    对建筑楼进行规划，切断开关和阀门的位置。
  - Inventory of holdings, with priorities for salvage marked on floor-plans.
    在建筑平面图上标记出重点救助的持有库存。
  - List of names, addresses, and home telephone numbers of personnel with emergency responsibilities.
    紧急责任人员的名称、地址和家庭电话号码的清单。
  - List of names, addresses, and home telephone numbers of the in-house disaster response team.
    内部的灾难救助团队的姓名、地址和家庭电话号码的清单。
  - List of names, addresses and home telephone numbers of trained conservators with experience in salvaging water-damaged materials, resource organizations, and

other facilities able to offer support in the event of a disaster.

有经验并训练有素的保护者的姓名、地址、家庭电话、打捞材料、资源组织和在其他灾难中提供支持的设备清单。

- List of disaster control services, in-house supplies and equipment, and in any central store, including locations and names of contacts with home telephone numbers.

  灾难控制服务、内部物资和设备清单，包括每个中心商场的地点和家庭电话号码、联系名称。

- List of suppliers of services and sources of additional equipment and supplies, including names of contacts and home telephone numbers.

  供给服务、额外设备和用品的清单，包括联系人的名字和家里的电话号码。

* Arrangements made to access freezing facilities.

  安排可用的冷冻设施。

* Arrangements for funding emergency needs.

  安排紧急资金的需求。

* Copies of insurance policies.

  保单副本。

* Salvage procedures.

  救助程序。

* Distribute the plan and documentation to appropriate locations on- and off-site.

  分发计划并将计划和文档分配到和场内场外适当的位置上。

Institute procedures to notify appropriate people of the disaster and assemble them rapidly. Therefore, disaster preparedness and management (for natural and man-made hazards) should be an essential part of any destinations' integrated management plan. Heat waves (such as that experienced in the European summer of 2003), the Southeast Asia tsunami of December 2004, changes in tropical storm intensity (such as Hurricane Katrina in August 2005) and forest f res (such as those experienced in Portugal and Greece in 2005) are all examples of how disasters and environmental emergencies can impact tourist destinations and holiday experiences.

程序调查，向合适的人通知有关灾情，并迅速将他们召集。因此，灾害防备和管理(自然灾害和人为灾害)应该是综合管理计划中任何目的的重要组成部分。高温(如2003年夏天的欧洲)、2004年12月东南亚海啸、热带风暴强度的变化(如2005年8月卡特里娜飓风)和森林大火(如2005年在葡萄牙和希腊的经历)，都是灾害和环境危机对旅游地和度假经历有很大影响的例子。

There have always been, and always will be disasters. The patterns of modern life are exposing more communities to danger then ever before. Tourist areas will be more and more exposed to rising sea levels, and meteorological-related disasters due to climate change. Disaster risk eduction linked with efforts to climate change adaptation and improvement of living conditions are undoubtedly today's major global challenges.

# Disaster Prevention and Reduction

一直以来，灾难始终相伴。现代生活的模式和灾难产生前所未有的交集。旅游区越来越多地暴露于上升的海平面。由于气候变化，又产生了与气象相关的灾害。为减轻灾害风险、适应气候变化和改善生活条件而努力，无疑是今天重大的全球性挑战。

Disasters inevitably bring about crises. It is the degree to which people are prepared for disaster that determines how vulnerable or resilient their community will be.

灾害必然带来危机。人们在何种程度上准备应对灾难，决定了哪些是弱势群体和弹性群体。

Phase III: Response
第三阶段：响应

- When disaster strikes, follow established emergency procedures for raising the alarm, evacuating personnel and making the disaster site safe.
  当灾难来袭时，按照既定的应急程序报警、疏散人员和保持灾难现场的安全。
- Contact the leader of the disaster response team to direct and brief the trained salvage personnel. When permission is given to re-enter the site, make a preliminary assessment of the extent of the damage, and the equipment, supplies and services required.
  联系救灾团的主要负责人，指引训练有素的打捞人员并简要说明情况。当被允许重新进入该地时，应对破坏程度作初步评估，并需要有设备、用品及服务。
- Stabilize the environment to prevent the growth of mould. Photograph damaged materials for insurance claim purposes.
  稳定环境，以防止霉菌的生长。对损失情况拍照以作保险索赔之用。
- Set up an area for recording and packing material which requires freezing, and an area for air drying slightly wet material and other minor treatment.
  为记录和包装材料设置了地方，这些材料需要干燥和其他的二次处理。
- Transport water-damaged items to the nearest available freezing facility.
  将水危害项目转移到最近的可用的冷冻设施。

Phase IV: Recovery
第四阶段：恢复

- Getting back to normal.
  恢复正常。
- Establish a programme to restore both the disaster site and the damaged materials to a stable and usable condition.
  建立一个计划，将灾难现场和损坏的材料恢复到稳定和可用的状态。
- Determine priorities for restoration work and seek the advice of a conservator as to the best methods and options, and obtain cost estimates.
  确定恢复工作的重点，考虑相关人员的意见以获得最好的灾后重建的方法，并得出成本估算结果。

## New Words and Expressions   生词和短语

1. cyclone   *n.* 旋风；[气象] 气旋；飓风
2. tsunamis   *n.* 海啸；津波(tsunami 的复数)
3. degradation   *n.* 退化；降格，降级；堕落
4. vermin infestation   害虫侵扰
5. volcanic   *n.* 火山岩
6. meteorological   *adj.* 气象的；气象学的
7. concentration   *n.* 浓度；集中；浓缩
8. casualty   *n.* 意外事故；伤亡人员
9. instability   *n.* 不稳定(性)；基础薄弱；不安定
10. leaking roof   屋顶漏水
11. blocked drain   堵塞的下水道
12. explosion   *n.* 爆炸；爆发；激增
13. spill   *n.* 溢出，溅出；溢出量
14. deficiency   *n.* 缺陷，缺点；缺乏
15. vibration   *n.* 振动；犹豫；心灵感应
16. abnormality   *n.* 异常；畸形
17. sprout   *n.* 苗；芽，幼芽；嫩枝
18. blossom   *n.* 花；开花期；兴旺期；花开的状态
19. landslide   *n.* [地质] 山崩；大胜利
20. collapse   *n.* 倒塌；失败；衰竭
21. debris flow   泥石流
22. turbid   *adj.* 浑浊的；混乱的；雾重的
23. bump   *n.* 肿块，隆起物；撞击
24. vapor   *n.* 蒸汽；烟雾
25. snowmelt   *n.* [水文] 融雪水
26. ditch   *n.* 沟渠；壕沟
27. evacuation   *n.* 疏散；撤离；排泄
28. backwater   *n.* 回水；死水；停滞不进的状态或地方
29. reservoir   *n.* 水库；蓄水池
30. ridges   *n.* 带钢单向皱纹
31. sewerage   *n.* 污水；排水设备
32. manuscript   *n.* 手稿；原稿
33. shelving   *n.* 倾斜，架子；搁置
34. water-sensing alarm   水传感报警
35. restoration   *n.* 恢复；复位；王政复辟；归还
36. resilient   *adj.* 弹回的，有弹力的
37. mould   *n.* 模具

# Chapter 15 Disaster Prevention and Reduction

## Exercises 练习

Ⅰ. Write a T in front of a statement if it is true according to the text and write an F if it is false.

1. Disaster refers to all the occurrences that can cause destructive influence to the safety of people's life and property and the environment that they rely on to survive.

2. Natural disasters can be prevented, but measures also should be taken to eliminate or reduce the possibility of trouble.

3. Collapse refers to the geological disaster that the soil or rock body on a steeper slope suddenly separates from the mountain due to gravity.

4. Before the landslide begins, animals within the scope of the landslide will fall in great panic.

5. Landslides are likely to happen during the snowmelt period in the winter each year.

6. Be alert when there are constant heavy rains and storms and the water level rises.

Ⅱ. Complete the following sentences.

1. Please list five nature disasters:_____.

2. _____ refers to the vibration of the surface of the earth caused by the sudden release of the energy slowly accumulated in the interior of the earth.

3. _____ refers to the geological disaster frequent in mountainous regions that floods flow along a slope with large amounts of mud and rocks.

4. How to escape at the time of landslide in a mountain?
_____
_____

5. _____ refers to rains of huge precipitation. It often causes flooding and serious soil and water losses.

6. Disaster Plan usually involves four phases :_____.

Ⅲ. Translate the following sentences into Chinese.

1. Growing populations, environmental degradation and global warming are making the impacts worse, creating greater disasters and making the need to find better ways to protect people more urgent.

2. Water flow suddenly discontinues in a river or a flood suddenly rises with large amounts of floating grasses and trees.

3. To prevent minor waterlogging inside a house, place water baffles or pile up earth ridges in front of the door according to local conditions.

4. If you have been buried under the ruins for a long time and the rescue personnel have not come or heard your signals for help, try your best to survive.

5. Tourist areas will be more and more exposed to rising sea levels, and meteorological-related disasters due to climate change.

6. When permission is given to re-enter the site, make a preliminary assessment of the extent of the damage, and the equipment, supplies and services required.

Ⅳ. Translate the following sentences into English.

1. 无论灾难是何种形式，其所带来的实际伤害通常是由水或火引起的。
2. 它们可能会大面积地发芽、开花、结出果实、枯萎并在不适当的季节变得非常繁茂。
3. 泥石流通常是突然的、凶猛的、快速的和短时间内发生的。
4. 滑坡停止后你不应该回家立即检查滑坡的情况。
5. 提前移动到一个安全的位置是抵御滑坡的最好方式。
6. 不乱扔垃圾，不把垃圾扔在道路上的排水口处，以避免造成堵塞和洪水。

# Answers　答案

Ⅰ.

1. T  2. F  3. T  4. T  5. F  6. T

Ⅱ.

1. Rain and wind storms; Floods; Biological agents; Earthquakes; Volcanic eruptions

2. Earthquake

3. Debris flow

4. i. Hide under a strong obstruction or crouch in a ditch or trench on the ground

　ii. Protect your head well and use your clothes to wrap your head up at time of landslide

5. Rainstorm

6. Prevention; Preparedness; Response; Recovery

Ⅲ.

1. 人口增长、环境恶化和全球变暖正在使影响变得更加严重，产生了更大的灾难，迫切需要寻找更好的方法来保护人们。
2. 水流量在河流中突然停止或夹带大量漂浮的草和树的洪水突然升起。
3. 为了防止室内小涝，根据当地的条件将挡水挡板或堆土埂在门前。
4. 如果你被掩埋在废墟里的时间很长，救援人员还没到或没听到你的呼救信号，尽你所能去活下来。
5. 由于气候的变化，旅游区将越来越受到海平面上升和气象灾害的影响。
6. 当被允许进入现场时，做一个对损伤程度的初步评估并列一个必需的设备、用品和服务的清单。

IV.

1. Regardless of the many forms a disaster may take, the actual damage to collections is usually caused by fire or water.

2. They may sprout, blossom, grow fruits, wither in large areas and become greatly thriving in an unduly season.

3. Debris flows are usually abrupt, ferocious, fast-moving and short in time.

4. You should not go home to check the situation immediately after the landslide stops.

5. Moving to a safe location in advance is the best way to defend against a landslide.

6. Do not throw garbage and litter into sewerage openings on the road to avoid causing blockage and flooding.

# Chapter 16 Planning, Scheduling and Construction Management
# 第 16 章 项目规划、组织及施工控制

The construction industry plays a significant role in the development of national economy. Almost half of the total outlay in any five year plan is utilized for construction activities which constitute an integral part of all development projects. During the last four decades, the construction industry in China has undergone large scale mechanization with rapid changes and advancement in construction practices as well as in the management of construction works. The term "Construction" is no longer limited only to the physical activities involving men, materials and machinery but covers the entire gamut of activities from conception to realization of a construction project.

建筑行业对国民经济发展起着至关重要的作用。在任何一个五年计划的总支出中，近一半额度用于建筑行业，而建筑行业是所有规划项目系统内不可或缺的一个部分。在过去的 40 年里，中国建筑业经历了大规模的机械化实践，在建设工程领域获得了快速的、长足的发展。"施工"不再是只涉及人力、材料及机械的活动，而是涵盖从构思到实现整个项目的全过程。

## 16.1 Construction Management 施工管理

Management is the science and art of planning, organizing, leading and controlling the work of organization members and of using all available organization resources to reach stated organizational goals.

施工管理是一门涵盖对项目各个部门的统一规划、组织、领导和管理，利用一切可利用的资源实现预先设定的目标的科学和艺术。

Construction Management includes review of contracts, order materials, hire and schedule sub-contractors, resource planning and to provide quality control and insure that the construction project is completed on time and with allocated budget. Planning and scheduling of activities involved in the construction process is extremely important to successfully achieve the goals of a construction project. A successful construction manager should have good leadership skills, effective communication skills and sound organization skills with knowledge of dispute resolution techniques. He should be able to motivate the construction team. Good knowledge of accounting

# Planning, Scheduling and Construction Management

and finance and quantity survey is also required.

建设管理包括合同审查、材料订购、确定分包商、资源规划以及确定工程质量并按时完工且不超预算所应采取的措施。在施工过程中涉及的分项工程的规划及有效的组织对保证建设项目目标如期完成是极其重要的。一个优秀的建筑施工负责人应具有良好的领导能力、有效的沟通能力及组织能力来解决问题。他应该能够激励施工团队，精通财务及质量管理的能力也是必需的。

Construction management deals with economical consumption of the resources available in the least possible time for successful completion of construction project. "Men", "materials", "machinery" and "money" are termed as resources in construction management.

良好的施工管理可以实现在最短的时间内、利用最小限度的资源消耗来圆满完成建设项目。"劳动力""材料""机械"和"资金"被称作施工管理过程中的资源。

## 1. Objectives of Construction Management 项目管理的目的

The main objectives of construction management are: Completing the work with in estimated budget and specified time; Maintaining a reputation for high quality workmanship; Taking sound decisions and delegation of authority; Developing an organization that works as a team.

项目管理的主要目的是：根据预算和所需时间完成工作；依靠优秀的质量保持声誉；根据权威指导做出正确的决定；将组织发展成一个团队。

## 2. Functions of Construction Management 项目管理的作用

The functions of construction management are：

管理的作用是：

(1) Planning: It is the process of selecting a particular method and the order of work to be adopted for a project from all the possible ways and sequences in which it could be done. It essentially covers the aspects of "What to do" and "How to do it", as shown in Figure 16.1.

规划：所谓项目规划是通过从众多可能的方法中选出一种可行的方法及施工进程来完成项目的实施。包括"做什么"和"如何做"等方面，如图16.1所示。

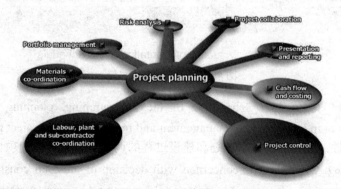

Figure 16.1　Project Planning

图 16.1　项目管理

Importance of planning:

规划的重要性：

① Planning helps to minimize the cost by optimum utilization of available resources.

规划最有效的可用资源，降低工程造价。

② Planning reduces irrational approaches, duplication of works and inter departmental conflicts.

规划可减少不恰当方法的使用、大量的重复工作及跨部门间的冲突。

③ Planning encourages innovation and creativity among the construction managers.

规划可激发施工管理人员的创新和创造能力。

④ Planning imparts competitive strength to the enterprise.

规划可赋予企业竞争实力。

(2) Scheduling: Scheduling is the fitting of the final work plan to a time-scale. It shows the duration and order of various construction activities. It deals with the aspect of "when to do it". Visual representation of the schedule is shown in Figure 16.2.

调度：调度是将施工进程与时间联系起来。它表明各阶段建设活动的持续时间和施工顺序，解决"什么时候做什么"的问题。调度的图表显示如图 16.2 所示。

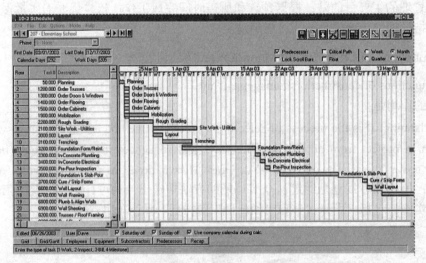

Figure 16.2　Visual Representation of the Schedule

图 16.2　调度的图表显示

Importance of scheduling: Scheduling of the programming, planning and construction process is a vital tool in both the daily management and reporting of the project progress.

调度的重要性：项目的规划、计划和建设过程的调度对日常管理和项目进展至关重要。

(3) Organizing: Organizing is concerned with decision of the total construction work into manageable departments/sections and systematically managing various operations by delegating specific tasks to individuals.

组织：组织通过将特定的任务分配给不同部门，将所有工作的决定传达至管理部门并进行系统的管理。

# Planning, Scheduling and Construction Management

(4) Staffing: Staffing is the provision of right people to each section /department created for successful completion of a construction project.

人员编制：指进行合理的人员安排，使一个建设项目得以顺利地完成。

(5) Directing: It is concerned with training sub ordinates to carryout assigned tasks, supervising their work and guiding their efforts. It also involves motivating staff to achieve desired results.

指导：通常是训练员工来完成既定的工作、监督实施效果并给予指导，也包括通过激励员工达到预期的结果。

(6) Controlling: It involves a constant review of the work plan to check on actual achievements and to discover and rectify deviation through appropriate corrective measures.

控制管理：指不断地回顾工作计划来检查实际运行的效果，通过适当的补救措施来发现并纠正工作方向。

(7) Coordinating: It involves bringing together and coordinating the work of various departments and sections so as to have good communication. It is necessary for each section to aware of its role and the assistance to be expected from others.

协调：它涉及召集和协调各部门和部分的工作，以便有良好的沟通。每个部门需要意识到自己的角色及期望从其他部门所获得的帮助。

## 3. Importance of Construction Management  施工管理的重要性

(1) Proper management practices invariably lead to "maximum production at least cost". A good construction management, results in completion of a construction project within the stipulated budget.

适当的管理措施，总是可以达到"以最少的成本获得最大的效益"的效果。一个好的施工管理，可使项目在预算中得以很好地完成。

(2) Construction management provides importance for optimum utilization of resources. In other words, it results in completion of a construction project with judicious use of available resources.

良好的施工管理可保证资源的最佳利用。换言之，可明智地利用现有资源来完成建设项目。

(3) Construction management provides necessary leadership, motivates employees to complete the difficult tasks well in time and extracts potential talents of its employees.

良好的施工管理可提高领导能力、激励员工及时高效地完成高难度的任务，并从中发掘有潜力的人才。

(4) Construction management is beneficial to society as the effective and efficient management of construction projects will avoid, escalation of costs, time overrun, wastage of resources, unlawful exploitation of labor and pollution of environment.

良好的施工管理有益于社会，因为高效的项目管理能够避免超出预算、工期延长、资源浪费、非法剥削劳工和环境污染。

## 16.2 Organizing for Project Management
## 项目管理组织

### 1. Project management　项目管理

The management of construction projects requires knowledge of modern management as well as an understanding of the design and construction process. Construction projects have a specific set of objectives and constraints such as a required time frame for completion. While the relevant technology, institutional arrangements or processes will differ, the management of such projects has much in common with the management of similar types of projects in other specialty or technology domains such as aerospace, pharmaceutical and energy developments.

建设项目的管理需要现代化管理知识以及对设计和施工过程有深入的理解。建设项目一般均有一定的目标和限制，比如说在规定的工期内完成。虽然相关技术、机构设置或过程会有所不同，但是这类项目的管理与其他相似类型的专业或技术领域有很多共同点，如航空航天、制药和能源发展项目的管理。

Generally, project management is distinguished from the general management of corporations by the mission-oriented nature of a project. A project organization will generally be terminated when the mission is accomplished. According to the Project Management Institute, the discipline of project management can be defined as follows:

一般来说，项目管理有别于发号命令式的企业管理模式。一个项目的组织通常会在任务完成时终止。根据项目管理协会的规定，项目管理可以定义如下：

Project management is the art of directing and coordinating human and material resources throughout the life of a project by using modern management techniques to achieve predetermined objectives of scope, cost, time, quality and participation satisfaction.

项目管理是艺术，通过现代管理技术，指导和协调整个项目生命周期中的人力和物力，实现预定的目标、范围、成本、时间、质量和参与满意度。

By contrast, the general management of business and industrial corporations assumes a broader outlook with greater continuity of operations. Nevertheless, there are sufficient similarities as well as differences between the two so that modern management techniques developed for general management may be adapted for project management.

与之对比，一般的商业和工业企业的综合管理应具有可连续性运行的更广阔的前景规划。尽管两者间有差异，但也有足够的相似性，因此用于一般商业管理的现代管理技术也可用于项目管理。

The basic ingredients for a project management framework may be represented schematically in Figure 16.3. A working knowledge of general management and familiarity with the special knowledge domain related to the project are indispensable. Supporting disciplines such as computer science and decision science may also play an important role. In fact, modern management practices and various special knowledge domains have absorbed various techniques or tools which were once identified only with the supporting disciplines. For example, computer-based information systems and decision support systems are now common-place tools for general management. Similarly, many operations research techniques such as linear programming and network analysis are now widely used in many knowledge or application domains. Hence, the representation in Figure 16.3 reflects only the sources from which the project management framework evolves.

项目管理框架的基本构成如图 16.3 所示。对一般性管理的手段与项目相关的特殊领域的了解是不可或缺的。支撑学科，如计算机科学和决策科学，会发挥重要作用。事实上，在现代管理实践和各种专业知识领域已经融合了各种曾被支撑学科识别的技术或手段。例如，以计算机为基础的信息系统和决策支持系统用于一般性管理是非常普遍的。同样，许多操作运行技术，如线性规划和网络分析研究现在被广泛地应用于许多知识或应用领域。因此，图 16.3 仅呈现了项目管理框架演变的来源。

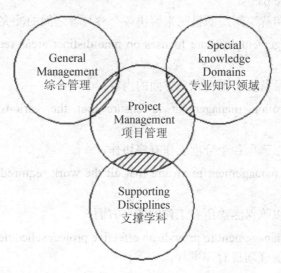

Figure 16.3  Basic Ingredients in Project Management
图 16.3  项目管理中的基本要素

Specifically, project management in construction encompasses a set of objectives which may be accomplished by implementing a series of operations subject to resource constraints. There are potential conflicts between the stated objectives with regard to scope, cost, time and quality, and the constraints imposed on human material and financial resources. These conflicts should be resolved at the onset of a project by making the necessary tradeoffs or creating new alternatives. Subsequently, the functions of project management for construction generally include the following:

具体来说，项目建设管理涵盖一系列通过实施一系列的受到资源限制的具体运行工作来完成既定目标。在既定的目标与项目规模、工期及质量控制、人力物力和财力的限制之间存在潜在的冲突。在项目开始时应该通过必要的权衡或提出新的方法手段来解决这些冲突。建设项目管理功能一般包括以下内容：

(1) Specification of project objectives and plans including delineation of scope, budgeting, scheduling, setting performance requirements, and selecting project participants.

项目的施工目标要求及涉及规模、预算、调度、性能要求以及选择项目参与者的规划。

(2) Maximization of efficient resource utilization through procurement of labor, materials and equipment according to the prescribed schedule and plan.

根据既定时间表和计划，通过雇佣劳动力、采购材料和设备，使资源消耗最小化。

(3) Implementation of various operations through proper coordination and control of planning, design, estimating, contracting and construction in the entire process.

通过全程中适当的协调和规划设计、评估、承包来落实执行各种工序。

(4) Development of effective communications and mechanisms for resolving conflicts among the various participants.

建立有效的沟通和解决纠纷的机制来解决各个参与者之间的冲突。

The Project Management Institute focuses on nine distinct areas requiring project manager knowledge and attention:

项目管理需要项目负责人重视九个方面的内容：

(1) Project integration management to ensure that the various project elements are effectively coordinated.

项目整合管理，以确保各个分项工作有序协作。

(2) Project scope management to ensure that all the work required (and only the required work) is included.

项目全局管理，以确保涵盖所有工作的各个方面。

(3) Project time management to provide an effective project schedule.

项目时间管理，保证项目有序进行。

(4) Project cost management to identify needed resources and maintain budget control.

项目成本管理，确定必需的投入及维护部分正常运行。

(5) Project quality management to ensure functional requirements are met.

项目质量管理，确保完成所需的功能要求。

(6) Project human resource management to development and effectively employ project personnel.

项目人力资源管理，保证发展及有效地使用项目人员。

(7) Project communications management to ensure effective internal and external communications.

项目信息管理，以确保有效的内部和外部沟通。

(8) Project risk management to analyze and mitigate potential risks.

项目风险管理，实现分析和化解潜在的风险。

(9) Project procurement management to obtain necessary resources from external sources.

项目采购管理，确保所需材料、资源的供应。

## 2. Trends in Modern Management　现代管理的发展趋势

In recent years, major developments in management reflect the acceptance to various degrees of the following elements:

近年来，项目管理科学的发展涵盖了以下几方面内容：

### 1) The Management Process Approach　管理过程控制办法

The management process approach emphasizes the systematic study of management by identifying management functions in an organization and then examining each in detail. There is general agreement regarding the functions of planning, organizing and controlling. A major tenet is that by analyzing management along functional lines, a framework can be constructed into which all new management activities can be placed. Thus, the manager's job is regarded as coordinating a process of interrelated functions, which are neither totally random nor rigidly predetermined, but are dynamic as the process evolves. Another tenet is that management principles can be derived from an intellectual analysis of management functions. By dividing the manager's job into functional components, principles based upon each function can be extracted.

Hence, management functions can be organized into a hierarchical structure designed to improve operational efficiency, such as the example of the organization for a manufacturing company shown in Figure 16.4. The basic management functions are performed by all managers, regardless of enterprise, activity or hierarchical levels. Finally, the development of a management philosophy results in helping the manager to establish relationships between human and material resources. The outcome of following an established philosophy of operation helps the manager win the support of the subordinates in achieving organizational objectives.

管理过程控制的方法是通过对一个项目组织过程中的管理识别功能检查管理的每个细节，强调系统性科学管理。这是对项目规划、组织和控制的过程所达成的普遍共识。一个主要的宗旨是，通过职能分析管理，可形成能够随时植入新的管理行为的一个框架。因此，管理者的工作职责可视为协调内部各功能间的相互关系，这既不是完全随机的，也非事先的硬性规定，而是涉及控制管理的动态特点。另一个宗旨是，管理原则可以通过对管理职能从理性的分析来获得。可将管理者工作分为功能模块及基于每个功能获得的管理理论模块。

因此，管理功能可概括为一个旨在提高运营效率的等级结构。如图16.4所示，一家制造公司组织。基本的管理职能由所有的管理者执行，与企业、活动或层次水平无关。最后，管理理念发展成为可帮助管理者建立人力和物力之间的关系的一门科学。

如下所建立的管理理念结果有助于实现组织目标，赢得客户的支持。

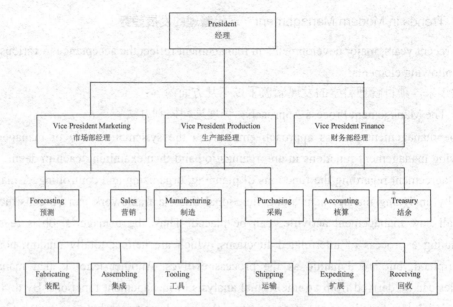

Figure 16.4　Illustrative Hierarchical Structure of Management Functions

图 16.4　管理功能的分层结构图

2) The Management Science and Decision Support Approach
管理科学和决策支持的方法

The management science and decision support approach contributes to the development of a body of quantitative methods designed to aid managers in making complex decisions related to operations and production. In decision support systems, emphasis is placed on providing managers with relevant information. In management science, a great deal of attention is given to defining objectives and constraints, and to constructing mathematical analysis models in solving complex problems of inventory, materials and production control, among others. A topic of major interest in management science is the maximization of profit, or in the absence of a workable model for the operation of the entire system, the suboptimization of the operations of its components. The optimization or suboptimization is often achieved by the use of operations research techniques, such as linear programming, quadratic programming, graph theory, queuing theory and Monte Carlo simulation. In addition to the increasing use of computers accompanied by the development of sophisticated mathematical models and information systems, management science and decision support systems have played an important role by looking more carefully at problem inputs and relationships and by promoting goal formulation and measurement of performance. Artificial intelligence has also begun to be applied to provide decision support systems for solving ill-structured problems in management.

管理科学与决策支持的方法有助于帮助管理者对相关运营及生产活动作出复杂的决策的定量方法的发展。决策支持系统的重点在于为管理人员提供相关的信息。在管理科学中，应投入大量的精力来决定目标和限制条件、构建数据模型解决有关库存、材料、生产控制的复杂问题。管理科学中的重大研究课题是确保利润的最大化，以及整个系统运行不具备

可行性时各部分的运作。优化过程通常使用运筹学技术，如线性规划、二次规划、图论、队论和蒙特卡罗模拟法等。此外，伴随着复杂的数学分析方法与信息系统发展的计算机技术的不断应用，通过系统分析输入的问题及其关系，管理科学及决策支持系统发挥了很大的作用。人工智能手段也已开始被应用到决策支持系统中来解决结构不良的管理问题。

3) The Behavioral Science Approach for Human Resource Development
人力资源行为科学的方法

The behavioral science approach for human resource development is important because management entails getting things done through the actions of people. An effective manager must understand the importance of human factors such as needs, drives, motivation, leadership, personality, behavior, and work groups. Within this context, some place more emphasis on interpersonal behavior which focuses on the individual and his/her motivations as a socio-psychological being; others emphasize more group behavior in recognition of the organized enterprise as a social organism, subject to all the attitudes, habits, pressures and conflicts of the cultural environment of people. The major contributions made by the behavioral scientists to the field of management include:

与人力资源有关的行为科学的方法非常重要，因为管理必须通过人的参与来完成任务。高效的管理者必须了解人力资源的重要性，如需求、动机、领导才能、人格、行为、团体合作等。在此背景下，需更注重基于个体性及动机的人际关系；作为社会有机体，对秩序井然的企业认知更强调群体行为，包括受到各种态度、习惯、压力和文化冲突的人。行为管理领域的科学家所做的重大贡献包括：

(1) The formulation of concepts and explanations about individual and group behavior in the organization.
制定组织中的个人和群体行为的概念和解释

(2) The empirical testing of these concepts methodically in many different experimental and field settings.
在许多不同的实验中和现场等处有条不紊地证实检验这些概念。

(3) The establishment of actual managerial policies and decisions for operation based on the conceptual and methodical frameworks.
在关于概念和方法框架的基础上的操作中，给出实际管理工作的政策和决定。

4) Sustainable Competitive Advantage  可持续的竞争优势

These four approaches complement each other in current practice, and provide a useful groundwork for project management.
这四种方法在目前的实践中相得益彰，并为项目管理提供了有益的基础。

Sustainable competitive advantage stems primarily from good management strategy. Project managers should be aware of the strategic position of their own organization and the other organizations involved in the project. The project manager faces the difficult task of trying to align the goals and strategies of these various organizations to accomplish the project goals. For example, the owner of an industrial project may define a strategic goal as being first to market with new products. In this case, facilities development must be oriented to fast-track, rapid construction. As another example, a contracting firm may see their strategic advantage in new technologies and emphasize profit opportunities from value engineering.

可持续的竞争优势主要来自良好的管理策略。项目管理者应该意识到其对本部门和参与项目的其他组织的战略作用。项目经理面临的艰巨任务是协调要求完成不同的目标和战略作用的各部门关系达到总体项目目标的实现。例如，一个工业项目的最高管理者做出一个抢先利用新产品来占领市场的战略定位。在这种情况下，生产设备必须加快发展、加快施工。另有一例，一个承包公司看到其在新技术的战略优势，根据项目可行性研究来强化工程收益。

### 3. Choice of Technology and Construction Method  技术及施工方法的选择

As in the development of appropriate alternatives for facility design, choices of appropriate technology and methods for construction are often ill-structured yet critical ingredients in the success of the project. For example, a decision whether to pump or to transport concrete in buckets will directly affect the cost and duration of tasks involved in building construction. A decision between these two alternatives should consider the relative costs, reliabilities, and availability of equipment for the two transport methods. Unfortunately, the exact implications of different methods depend upon numerous considerations for which information may be sketchy during the planning phase, such as the experience and expertise of workers or the particular underground condition at a site.

对工业设计、技术手段及施工方法的替代产品的研发过程往往是结构不良但又至关重要的。例如，决定是采用泵送还是桶运输混凝土将直接影响到施工建设的成本和工期。两个方案需考虑相对成本、可靠性和设备的可用性。但是，不同方法的确切实施取决于众多的因素，计划阶段要对实施过程中的各种信息进行研判，如在施工现场的特定条件下工人的经验和专业知识水平等。

In selecting among alternative methods and technologies, it may be necessary to formulate a number of construction plans based on alternative methods or assumptions. Once the full plan is available, then the cost, time and reliability impacts of the alternative approaches can be reviewed.

This examination of several alternatives is often made explicit in bidding competitions in which several alternative designs may be proposed or value engineering for alternative construction methods may be permitted. In this case, potential constructors may wish to prepare plans for each alternative design using the suggested construction method as well as to prepare plans for alternative construction methods which would be proposed as part of the value engineering process.

在选择可替代的方法和技术时，可能需要制订一定数量的基于替代方法或假设的施工计划。一旦完整计划是可行的，那么就可以对成本、时间和可靠性影响的替代方法进行相应的审查。

对几个备选方案进行验证在投标竞争中是明确的，可行设计方案及可选的施工方法均可得到验证。在这种情况下，具体的施工人员可对每一备选方案采用推荐的施工方法进行准备；反之，也可采用作为工程评价体系的一个部分的施工方法来进行计划准备。

# Chapter 16

## Planning, Scheduling and Construction Management

In forming a construction plan, a useful approach is to simulate the construction process either in the imagination of the planner or with a formal computer based simulation technique. By observing the result, comparisons among different plans or problems with the existing plan can be identified. For example, a decision to use a particular piece of equipment for an operation immediately leads to the question of whether or not there is sufficient access space for the equipment. Three dimensional geometric models in a computer aided design (CAD) system may be helpful in simulating space requirements for operations and for identifying any interferences. Similarly, problems in resource availability identified during the simulation of the construction process might be effectively forestalled by providing additional resources as part of the construction plan.

要形成一个施工计划，一个行之有效的方法就是根据项目实施者的设想或采用常规的计算机模拟技术来模拟施工过程。通过结果分析，可以分辨得出可选择的实施方案与采用方案的实施效果。例如，使用一个特定的设备进行操作可导致一个问题，即是否有足够的设备操作空间？在计算机辅助设计(CAD)系统中的三维几何模型有助于分析运行过程的操作空间、干扰识别。类似的，在施工模拟过程中，对资源分析时产生的问题可以通过采用其他符合施工方案的方法而有效地加以解决。

Example 1: A roadway rehabilitation
示例1：巷道修复

An example from a roadway rehabilitation project in Pittsburgh, can serve to illustrate the importance of good construction planning and the effect of technology choice. In this project, the decks on overpass bridges as well as the pavement on the highway itself were to be replaced. The initial construction plan was to work outward from each end of the overpass bridges while the highway surface was replaced below the bridges. As a result, access of equipment and concrete trucks to the overpass bridges was a considerable problem. However, the highway work could be staged so that each overpass bridge was accessible from below at prescribed times. By pumping concrete up to the overpass bridge deck from the highway below, costs were reduced and the work was accomplished much more quickly.

宾夕法尼亚州匹兹堡的一个巷道修复项目可以用来说明良好的建设规划和技术选择的重要性。在这个项目中，天桥桥面结构以及高速路面铺装将被取代。最初的计划是从两端的天桥向外建设，因此低于桥梁的公路表面被代替。由此，设备和混凝土卡车如何到天桥成为一个大问题。然而，高速公路施工是分期完成的，每个立交桥下面在规定的时期可被利用。通过从公路下面到天桥桥面泵送混凝土，可以降低工作成本，更快捷地完成工作任务。

## 16.3 Defining Work Tasks 规定工作任务

At the same time that the choice of technology and general method are considered, a parallel step in the planning process is to define the various work tasks that must be accomplished. These work tasks represent the necessary framework to permit scheduling of construction activities, along with estimating the resources required by the individual work tasks, and any necessary precedence or required sequence among the tasks. The terms work "tasks" or "activities" are often used interchangeably in construction plans to refer to specific, defined items of work. In job shop or manufacturing terminology, a project would be called a "job" and an activity called an "operation", but the sense of the terms is equivalent. The scheduling problem is to determine an appropriate set of activity start time, resource allocations and completion times that will result in completion of the project in a timely and efficient fashion. Construction planning is the necessary fore-runner to scheduling. In this planning, defining work tasks, technology and construction method is typically done either simultaneously or in a series of iterations.

在选择技术和采用的方法的同时，在计划过程中与其并行的、需完成的工作就是确定在该过程中必须完成的大量的工作任务。连同单项工程所需的材料预算及任何工序的提前或滞后等，这些工作任务代表了整个框架任务来确保施工进度。"任务"或"活动"通常可以在具体参考定义的项目的施工计划中互换使用。在工作车间或制造行业，项目将被称为"工作"和生产活动被称为"运行"，在某种意义上是等价的。进度控制是确定一项生产活动的开始时间、资源分配和完工时间，这将决定项目是否能按时、有效地完成。计划是实施的必要前提。在制订计划过程中，明确工作任务、采用的技术和施工方法通常通过累计或一系列的迭代来完成。

The definition of appropriate work tasks can be a laborious and tedious process, yet it represents the necessary information for application of formal scheduling procedures. Since construction projects can involve thousands of individual work tasks, this definition phase can also be expensive and time consuming. Fortunately, many tasks may be repeated in different parts of the facility or past facility construction plans can be used as general models for new projects. For example, the tasks involved in the construction of a building floor may be repeated with only minor differences for each of the floors in the building. Also, standard definitions and nomenclatures for most tasks exist. As a result, the individual planner defining work tasks does not have to approach each facet of the project entirely from scratch.

确定适当的工作任务可能是一个辛苦而乏味的过程，但它代表项目实施的必要信息。由于施工项目可以包括成千上万的单项任务，所以这个阶段的工作是费时费力的。幸运的是，很多工作计划模型可在不同项目中重复利用。例如，建筑楼面的任务可用于该建筑中的任意楼层。同时，标准的定义和术语对于大多数任务都适用。因此，近似项目的单工作任务没有必要完全从头开始。

Example 2: Task Definition for a Road Building Project
示例2：道路建设项目任务

As an example of construction planning, suppose that we wish to develop a plan for a road construction project including two culverts. Initially, we divide project activities into three categories as shown in Figure 16.5: structures, roadway, and general. This division is based on the major types of design elements to be constructed. Within the roadway work, a further sub-division is into earthwork and pavement. Within these subdivisions, we identify clearing, excavation, filling and finishing (including seeding and sodding) associated with earthwork, and we define watering, compaction and paving sub-activities associated with pavement. Finally, we note that the roadway segment is fairly long, and so individual activities can be defined for different physical segments along the roadway path. In Figure 16.5, we divide each paving and earthwork activity into activities specific to each of two roadway segments. For the culvert construction, we define the sub-divisions of structural excavation, concreting, and reinforcing. Even more specifically, structural excavation is divided into excavation itself and the required backfill and compaction. Similarly, concreting is divided into placing concrete forms, pouring concrete, stripping forms, and curing the concrete. As a final step in the structural planning, detailed activities are defined for reinforcing each of the two culverts. General work activities are defined for move in, general supervision, and clean up. As a result of this planning, over thirty different detailed activities have been defined.

以建设规划为例，假设我们想开发一个有两个涵洞的规划道路建设项目。最初，我们将项目分为三个类别，如图16.5所示：结构、道路和一般环节。这种划分是基于所要施工的主要单元。在道路施工中，进一步划分为土方工程和路面工程。在此工作环节，确定有关土方工程的整平、开挖、回填、完工(包括种草和铺草皮)等；定义有关路面工程的浇水、压实、摊铺阶段。最后，应注意到，施工路段相当长，因此可沿道路长度划分不同单个施工段。在图16.5中，将路面工程和土方工程划分到每一施工段。对于涵洞施工，定义了结构开挖、混凝土施工及配筋。更具体地说，结构开挖分为开挖本身和所需回填及压实。同样，混凝土分为施工形式、混凝土浇注和混凝土养护。该结构操作的最后一步是对两个涵洞的结构配筋。一般工作环节指为工作收尾、监督和施工现场清理。针对这个计划，共确定了30多个不同的操作活动。

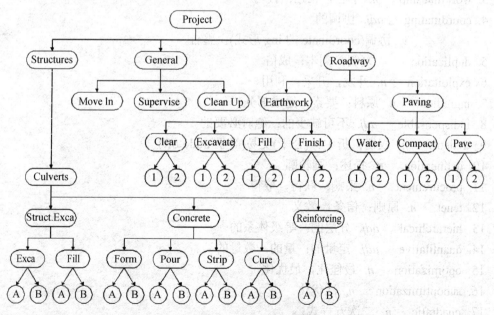

Figure 16.5　Illustrative Hierarchical Activity Divisions for a Roadway Project
图16.5　道路建设项目分层活动区域图

At the option of the planner, additional activities might also be defined for this project. For example, materials ordering or lane striping might be included as separate activities. It might also be the case that a planner would define a different hierarchy of work breakdowns than that shown in Figure16.5. For example, placing reinforcing might have been a sub-activity under concreting for culverts. One reason for separating reinforcement placement might be to emphasize the different material and resources required for this activity. Also, the division into separate roadway segments and culverts might have been introduced early in the hierarchy. With all these potential differences, the important aspect is to insure that all necessary activities are included somewhere in the final plan.

在项目的规划中，也可定义附加的一些生产活动。例如，材料采购或车道规划可作为单独的部分。项目的实施规划人员会定义一个不同于图 16.5 的层次分明的节点图。例如，浇注混凝土可作为涵洞施工的一个子项目。未考虑钢筋布置工序的一个原因可能是这项生产活动强调不同的材料和资源。同时，不同的道路施工段和涵洞施工可作为整个施工等级的较早引入部分。所有这些潜在的差异，重要的目的是确保不可或缺的施工工序被涵盖在整体施工计划中。

## New Words and Expressions  生词和短语

1. scheduling    *n.* 时序安排，行程安排；调度
2. mechanization    *n.* 机械化；机动化
3. workmanship    *n.* 手艺，工艺；技巧
4. coordinating    *adj.* 协调的
    *v.* 协调(coordinate 的 ing 形式)；整合
5. duplication    *n.* 复制；副本；成倍
6. exploitation    *n.* 开发，开采；利用
7. ingredient    *n.* 原料；要素；组成部分
8. indispensable    *adj.* 不可缺少的；绝对必要的
9. tradeoff    *n.* 权衡；折中；(公平)交易(等于 trade-off)
10. delineation    *n.* 描述；画轮廓
11. procurement    *n.* 采购；获得，取得
12. tenet    *n.* 原则；信条；教义
13. hierarchical    *adj.* 分层的；等级体系的
14. quantitative    *adj.* 定量的；量的，数量的
15. optimization    *n.* 最佳化，最优化
16. suboptimization    *n.* 次优化
17. quadratic    *n.* 二次方程式；
    *adj.* [数] 二次的
18. queuing theory    排队论；研究等待时间的等候理论
19. formulation    *n.* 构想，规划；公式化；简洁陈述

# Planning, Scheduling and Construction Management

20. roadway    *n.* 道路；路面；车行道；铁路的路基
21. rehabilitation    *n.* 复原
22. truck    *n.* 卡车；交易；手推车；载重汽车
23. terminology    *n.* 术语，术语学；用辞
24. iteration    *n.* [数] 迭代；反复；重复
25. nomenclature    *n.* 命名法；术语
26. scratch    *n.* 擦伤；抓痕；刮擦声；乱写
27. excavation    *n.* 挖掘，发掘
28. backfill    *n.* 回填；用来填满坑穴的泥土或其他东西
       *vt.* 回填；装填
29. compaction    *n.* 压紧；精简；密封；凝结

# Exercises 练习

Ⅰ. Write a T in front of a statement if it is true according to the text and write an F if it is false.

1. The term "Construction" is limited only to the physical activities involving men, materials and machinery.

2. Construction management deals with economical consumption of the resources available in the least possible time for successful completion of construction project.

3. Planning could reduce irrational approaches, duplication of works and inter departmental conflicts.

4. Computer-based information systems and decision support systems are now common-place tools for general management.

5. The definition of appropriate work tasks cannot represent the necessary information for application of formal scheduling procedures.

Ⅱ. Complete the following sentences.

1. Please list five functions of construction management : _____.

2. _____ involves bringing together and coordinating the work of various departments and sections so as to have good communication.

3. Maximization of efficient resource utilization through procurement of _____ according to the prescribed schedule and plan.

4. The terms work _____ are often used interchangeably in construction plans to refer to specific, defined items of work.

5. In this defining the various work tasks planning, defining _____ is typically done either simultaneously or in a series of iterations.

Ⅲ. Translate the following sentences into Chinese.

1. During the last four decades, the construction industry in China has undergone large scale mechanization with rapid changes and advancement in construction practices as well as in the management of construction works.

2. Staffing is the provision of right people to each department created for successful completion of a construction project.

3. By contrast, the general management of business and industrial corporations assumes a broader outlook with greater continuity of operations.

4. The outcome of following an established philosophy of operation helps the manager win the support of the subordinates in achieving organizational objectives.

5. The behavioral science approach for human resource development is important because management entails getting things done through the actions of people.

Ⅳ. Translate the following sentences into English.

1. 建筑业在国民经济发展中起着重要的作用。
2. 对各部门来说认识到自己的角色和从别人那里获得帮助是必要的。
3. 建设项目的管理需要现代管理知识，以及对设计和施工过程的理解。
4. 另一个信条是管理原则来源于管理功能的智能分析。
5. 在决策支持系统方面，重点是提供管理人员的相关信息。

## Answers　答案

Ⅰ.

1. F　2. T　3. T　4. T　5. F

Ⅱ.

1. Planning；Scheduling；Organizing；Staffing；Directing
2. Coordinating
3. labor, materials and equipment
4. "tasks" or "activities"
5. work tasks, technology and construction method

Ⅲ.

1. 在过去的 4 年中，随着工程建筑和建筑工程管理的快速的变化和进步，中国建筑业已经经历了大型机械化作业的过程。

2. 员工分配规定把合适的人安排到能顺利完成建设项目的各个部门。

3. 相比之下，一般的商业管理和工业企业之间通过更大的业务连续性来对前景进行一个更广阔的假设。

4. 坚持经营哲学的结果是帮助管理人员在实现组织目标等方面赢得下属的支持。

5. 行为科学的方法对人力资源开发很重要，因为管理需要通过人们的行动进行日程管理。

Ⅳ.

1. The construction industry plays a significant role in the development of national economy.

2. It is necessary for each section to aware of its role and the assistance to be expected from others.

3. The management of construction projects requires knowledge of modern management as well as an understanding of the design and construction process.

4. Another tenet is that management principles can be derived from an intellectual analysis of management functions.

5. In decision support systems, emphasis is placed on providing managers with relevant information.

# 参 考 文 献

[1] 武秀丽. 土木工程专业英语[M]. 北京：中国铁道出版社，2000.

[2] 董祥. 土木工程专业英语[M]. 南京：东南大学出版社，2011.

[3] 霍俊芳. 土木工程专业英语[M]. 北京：北京大学出版社，2010.

[4] 白越. 土木工程专业英语[M]. 重庆：重庆大学出版社，2011.

[5] Structural Concepts and Systems for Architects, Victor.E. Saouma. 1997 (Lecture notes) University of Colorado.

[6] "Evaluation and Rehabilitation of Concrete Structures" Mexico City, September 11-13, 2002 ,Jay H. Paul, S.E., P.E.

[7] Construction of Masonry Building with Appropriated Technologies 2004, Japanese Advisor Committee. Lima Peru.